T0324832

QUALITY PROCEDURES
FOR
HARDWARE AND SOFTWARE

QUALITY PROCEDURES FOR HARDWARE AND SOFTWARE

A Cost Effective Guide to Establishing a Quality System—Contains Manuals and Template Procedures

DAVID J. SMITH

B.Sc., C.Eng., F.I.E.E., F.I.Q.A., F.Sa.R.S., M.I.Gas.E.

Tonbridge, Kent, UK

and

JOHN S. EDGE

B.A., C.Eng., M.B.C.S., M.I.Q.A., M.Sa.R.S., M.A.P.M.

Englefield Green, Surrey, UK

ELSEVIER APPLIED SCIENCE

LONDON and NEW YORK

ELSEVIER SCIENCE PUBLISHERS LTD
Crown House, Linton Road, Barking, Essex IG11 8JU, England

Sole Distributor in the USA and Canada
ELSEVIER SCIENCE PUBLISHING CO., INC.
655 Avenue of the Americas, New York, NY 10010, USA

WITH 9 TABLES and 46 ILLUSTRATIONS

© 1991 ELSEVIER SCIENCE PUBLISHERS LTD

British Library Cataloguing in Publication Data

Smith, David J.
 Quality procedures for hardware and software.
 1. Manufacturing industries. Quality control. Applications of
 computer systems.
 I. Title. II. Edge, John S.
 658.562 0285

 ISBN 1-85166-550-1

Library of Congress Cataloging-in-Publication Data

Smith, David J. (David John), 1943–
  Quality procedures for hardware and software: a cost effective
 guide to establishing a quality system; contains manuals and
 template procedures / by David J. Smith, John S. Edge.
   p. cm.
 Includes bibliographical references and index.
 ISBN 1-85166-550-1
  1. Computer industry—Quality control. I. Edge, John S.
 II. Title.
 TK7886.S65 1991
 004'.068'5—dc20   90-47370
                                                              CIP

Contents

Chapter 5 Other Working Documents

PART 3.   OTHER ESSENTIAL GUIDANCE

Chapter 6 The Overview of Quality

Chapter 7 Auditing the System

PART 4. SAMPLE QUALITY MANUALS

PART 5. SAMPLE PROCEDURES

Acknowledgements

The authors would like to thank Gordon Bairstow, quality management consultant, for his thorough study of the manuscript. His helpful suggestions in resolving a number of anomalies were of great help, as was his painstaking editing of the cross-references.

Our thanks are also due to Mike Pilditch for some helpful comments on Chapter 9.

I (D.J.S.) am so grateful to my late wife, Margaret, for her encouragement and considerable help with the word processing. Sadly, she died shortly before the completion of the project but her efforts will not be forgotten. I (J.S.E.) would like to thank my wife Alexandra and our children Felicity and Edmund for their considerable patience during the preparation of the manuscript.

Introduction

How to Use This Book

The primary purpose of this book is to assist small companies, involved in both hardware and software, to devise and evolve their own quality systems. There are a number of national and now international standards which outline the activities for which procedures and records need to be specified. They are described and compared in Chapter 2, and the subsequent guidance in the book is intended to assist in meeting them.

Although, at first sight, the operations of a hardware equipment developer may seem very different from those of a software house, the basic requirements of a quality system, such as the BS 5750 and ISO 1987 series of documents, are the same. For this reason the same standard can be called for in both areas and it will be seen, in Part 2, that suitable procedures can be derived to meet both types of operation.

Quality standards (BS 5750, AQAP, ISO 9000 series) distinguish between companies carrying out, on the one hand, both design and manufacturing functions and, on the other hand, those who only manufacture to fixed specifications. In practice, the lesser requirements (those applying to manufacture to fixed specifications) are common to both levels of standard and the additional controls pertaining to design are added to obtain the higher standard. Chapter 2 explains the differences in detail.

It is not a simple matter to define a 'small company'. However, this book is aimed at organisations who have identified the need for a formal quality system but have only limited expertise and resources in producing quality procedures and instructions. Thus companies with headcounts in the range of a few to tens of persons are likely to fit that description.

At the lower end of the scale many companies in the 4–15 headcount range have no formal test or quality procedures and appoint an existing director, or other senior member, to have part-time responsibility for quality. The individual may have no specific experience in quality

1

management and will be seeking some form of example or 'Model Quality System' in order to instal a system from scratch. On the other hand, some slightly larger companies may already have a number of written inspection/ test procedures and, perhaps, a draft quality manual but are aware that their system falls short of current published standards. It is these two types of company for which this book has been written.

Part 1 explains the need for and the types of standards in use.

Part 2 describes the requirements in detail and provides guidance on how to select from the template procedures in Parts 4 and 5. It is laid out using the same headings as BS 5750 Part 1 1987 and for each a sample procedure is provided for:

Company A—a small electrical development and manufacturing company.

Company B—a small hardware and software systems development house.

Company C—a vendor of software packages who also offers support.

Part 3 addresses important subsequent activities such as:

—Auditing the system
—Collecting and optimising costs
—Obtaining approvals

Parts 4 and 5 provide a set of template manuals and procedures which the user may copy or amend to fit his own operation.

A number of consultants are available who will prepare and develop quality systems and who will, in addition, audit their implementation and assist in resolving the remedial actions which arise from such audits. Such consultancy is, in many cases, a cost-effective way of acquiring such expertise, particularly where the size of company does not justify a full-time quality professional. Nevertheless, the drafting and formatting of the actual procedures and work standards must involve an input from the company equal to at least the same amount of time as from its consultant.

The aim of this book, therefore, is to enable a company to proceed to the level of a draft quality system before investing in consultancy effort to assist in its implementation and development.

It is intended that, having read and considered Parts 1 and 2 of this book, Parts 4 and 5 be consulted in order to obtain a template or, at least, example procedures. The examples may be used and/or amended as necessary to fit the particular business in question, and guidance is given, in Chapter 4, as to the scope and purpose of each area of requirements.

PART 1

The What and Why of Quality Systems

Chapter 1

Why Quality Systems?

1.1 WHY ARE QUALITY SYSTEMS STANDARDS NECESSARY?

Manufacturing organisations survive only by selling their goods. Competition has never been more fierce and customers are increasingly seeking the *best value for money.*

In the days of the one-man business it was relatively easy to relate the customer's expectations to the designer, the producer and to the tester—all of these roles being filled by the same person whose interests were one and the same, namely the success of the business. As these small businesses grew the need for organisational relationships and disciplines was recognised. And so functional specialisations evolved as, for example, salesmen, engineers and accountants. With this came the management structure to assist communications and to control the operation.

A number of stages of development can be traced over the years:

—craftsmen
—standards imposed by foremen
—independent inspection
—calibration
—statistical methods
—zero defects approaches

Inspection alone is, however, not sufficient. The entire manufacturing operation must be committed to meeting the quality needs of the design such that every employee understands that quality has to be built in to the product. The following four golden rules apply equally to hardware and software:

—Goods must be designed to meet the customer's needs and in such a way that manufacture and maintenance are easy.

—Goods must be made consistently to the specified design and to the tolerances laid down.

—Marketing must ensure accurate advertising, punctual delivery, efficient servicing and effective market research that feeds back into continuing design improvements.

—There must, above all, be a total and organised commitment to quality from the top of the organisation.

Quality management systems are required to implement this philosophy throughout the organisation in a consistent ongoing manner. The management system must, therefore, be expressed as a written standard such as to provide a reference point and to permit objective measurement. This type of control is best achieved by means of a defined quality system standard.

1.2 WHAT IS A QUALITY SYSTEM?

There have been several attempts to define 'quality'. Each depends upon a particular viewpoint, such as that of the user, manufacturer, designer or salesman. Examples are:

Conformance to the Requirements (Crosby)

This is a manufacturing view of quality.

Fitness for Use (Juran)

and

The totality of features and characteristics of a product that bear on its ability to satisfy a given need (American and European quality societies)

These are marketing views of the same concept.

The quality of a product is the degree of conformance of all the relevant features and characteristics of the product to all of the aspects of a customer's need limited by the price which he is willing to pay (Groocock)

This latter definition is an attempt by John Groocock to imply that quality can be expressed as a number over a range and allows the evaluation of requirements, design, parts, manufacture and use of a product.

The common factor in these definitions is that they describe a requirement to conform to some specific features or attributes. Conformance, in turn, implies standardisation, which in practice calls for precisely documented requirements, standards, practices and procedures.

A major reason for written procedures and practices is that, unlike computers, the human brain is not well adapted to accurate repetition. Hence standardised rules substantially reduce (although do not entirely eliminate) the probability of error. If accurate repetition by one person is difficult to assure then standardisation, involving a number of persons carrying out the same task, is very much more so.

In order to consistently design or manufacture items, or for that matter provide services, against defined standards, then a formal arrangement of procedures and working practices is needed. Human beings are creative and tend to exercise ingenuity in the course of their activities. Important as these attributes are, they are not always best suited to consistent design and manufacture, and a disciplined adherence to methods is essential.

Such documentation covers three areas:

—*Descriptions of the product or service:* material descriptions, drawings, specifications and software listings which define the attributes to be reproduced.

—*The design, manufacturing and materials control procedures:* these define the order in which tasks are to be performed, by whom they are carried out, the records to be maintained, the nature of checks and reviews to be carried out, and so on.

—*Work standards and instructions:* more specific than the above procedures, work standards and instructions provide explicit guidance in carrying out particular tasks. Examples are drafting practice, soldering methods, test and inspection procedures, and housekeeping.

It is the latter two groups which together constitute a quality management system, the aim of which is to control and monitor the first group (product description). The quality management system provides a means of monitoring activities and processes, and, wherever necessary, carrying out remedial actions.

Originally, quality systems evolved in engineering and manufacturing situations as a result of the realisation that the inherent variability in repetitive production needed to be controlled. In the 1970s this concept spread to the design activity, as it became evident that the same principles of standardisation and review could be applied. In the 1980s two other areas of activity began to benefit from the application of quality methodology.

One such area is the provision of services, to which the principles of defined performance criteria, together with procedures and audits to ensure conformance, have been successfully applied. The other is the design and maintenance of software. Software includes both real time applications programs, which often form part of hardware engineering products, and batch software, which provides off-line calculation, data processing, accounting and other functions.

Written procedures are a snapshot of the actual activities in a company at a particular point in time. They should, therefore, evolve and be subject to regular review and change to meet the changing needs of the business. Issue numbers and proper documentation control, as applied to all the other specifications and drawings, is therefore essential.

It should always be kept in mind that a quality system exists to serve the business. It should meet the needs of the particular development, software or manufacturing operation and should aid optimum cost-effective quality to be achieved. It is a mistake to impose a system over and above the best ways of working. It should be tailored to blend in with the appropriate procedures and work standards necessary to run the business correctly.

The standards (BS 5750 and ISO 9000) do not therefore demand any specific procedure, only that adequate procedures do exist and that they are implemented in the areas stated. Nor do they imply any predetermined quantity of procedures. A very small organisation might conceivably satisfy the requirements with a single quality manual containing all the essential procedures. A larger company, on the other hand, will likely need a number of separate procedures and a quality manual acting as a top level summary document. Chapter 3 elaborates on the hierarchy of documents required.

Quality procedures and manuals are not for show but for day-to-day use, and should therefore be concise and readable. Well-used copies should be much in evidence. A single pristine set in each manager's office is evidence that little of the system is being implemented.

1.3 HOW IS A QUALITY SYSTEM ACQUIRED?

From very early times, much of the civil, technical and industrial progress we have made has been initiated as a response to or as a consequence of military needs. Originally, reliability, rather than cost, has been the prime factor in setting military requirements and as a result military procurement evolved formalised inspection systems.

Thus suppliers of military equipment, as a follow-on from inspection procedures, acquired quality management systems. Inspection alone, however, is not sufficient to ensure the quality of complex systems and therefore the requirements of BS 5750 (which grew out of the military 05-21 series of standards) embraces all aspects of the business from conceptual design through procurement, manufacture and test to installation and maintenance. Military quality requirements extend not only to prime contractors but to their subcontractors through common procedures and standards which cover all aspects of design and manufacture.

In the private sector, particularly the large multi-national companies, the need for quality procedures and standards has always received attention. These large organisations are characterised by multi-site operations producing a wide range of products.

These circumstances demand, from a commercial viewpoint, consistency of methods, procedures and standards. The implementation of quality systems in large companies soon highlighted the existence of quality-related costs (see Chapter 8) and the savings that could be made as a result of effective controls.

It was realised that quality-related costs were not wholly determined by, or controlled by, the company but also by suppliers, subcontractors, stockists, dealers and agents. The real penalty costs arising from poor materials, unusable components and shoddy subassemblies could greatly exceed their original purchase costs.

Thus large organisations, with purchasing power, started to demand that their suppliers implemented quality procedures and standards. As a result, smaller companies (the suppliers and subcontractors of the large companies) started to take quality more seriously. It was in many cases their first exposure to the need for formalised quality control. In some cases these quality systems were perceived as a necessary evil; necessary in order to do business with the larger company rather than as a positive opportunity to improve the management control and financial health of the operation. As a business parameter, quality was only dimly understood.

With increasing national and international competition, quality is becoming more accepted and understood as an essential business parameter, which is needed both to stay in the market and to produce goods cost-effectively. It should be a prime requirement of any company that it meets the needs of potential and current customers. If it does not, they will likely take their custom elsewhere. In order to explore new markets, and for that matter to retain existing ones, it is necessary to put the quality requirements of customers first.

Today both public and private sector purchasers (as well as consumers) increasingly demand rather than expect quality of both products and services. As a result, common quality procedures and standards have been passed down from the larger organisations, through the hierarchy of subcontract and supply, to the many medium and small companies with whom they do business. It is for those smaller companies for whom this book has been written.

An additional incentive, which involves safety, has arisen from the Consumer Protection Act of 1987. Products must be designed, manufactured and maintained so as to avoid death or personal injury, and the 1987 Act allows few defences on the part of anyone in the design and manufacturing chain. Liability is absolute.

Another incentive will be the European Market situation of 1992, as a result of which competition will increase. Those with a proven quality system, who will thus have optimum quality costs, will stand a far better chance of succeeding and surviving in that situation.

The following chapters review the current published standards which specify frameworks for creating a quality management system. They then provide guidance on how to develop one's own quality system and provide numerous sample manuals, procedures and work instructions which can be copied or modified accordingly.

Chapter 2

Comparison of Current Standards

2.1 THE EVOLUTION OF MOD AND ISO STANDARDS

The terms BSI, MOD, ISO and LRQA stand for British Standards Institute, Ministry of Defence, International Standards Organisation and Lloyd's Register Quality Assurance.

The following is a brief history of events.

1960: NATO (North Atlantic Treaty Organisation) produced the Allied Quality Assurance Publications (AQAP procedures) to ensure uniform standards in procured supplies. The principal elements of a quality assurance management system were defined. There was a shift of emphasis from approved inspection to:

(a) direct and complete control of quality by the supplier;
(b) audit surveillance from the purchaser.

1971: BS 4778 was produced by BSI. This standard defined the terminology used in quality assurance and was published to reduce the confusion which existed around the meaning and application of quality terminology.

BS 4778 has remained in use since this data and has now been harmonised internationally by the publication of ISO 8402; dual numbered in the UK as BS 4778/ISO 8402.

1972: BS 4891 was produced as a guide to quality assurance. This was intended to assist the supplier and purchasing (assessment) organisations alike in understanding quality principles. As a guide it has no contractual status.

The UK Ministry of Defence, in attempting to introduce the AQAPs into its defence procurement, recognised the need to

11

'Anglicise' the AQAPs and provide guidance to suppliers and its own assessors. In providing Defence Standards 05-21 to 05-29 it laid down foundations for everything which has so far developed. DEF STANs 05-21, 05-24 and 05-29 were contractually enforceable.

1974: Under pressure from commercial industry, which was also heavily involved in defence work, BSI produced BS 5719, which established guidelines for Quality Systems for application in the 'non-military' field. This standard was widely adopted by the nationalised industries, who used it for all major procurement. In parallel, a committee, chaired by Frederick Warner (now Sir), was examining the whole subject of harmonisation of standards—including those for quality.

1976: The Warner Committee report recommended, amongst others, a UK standard for quality systems to be accepted 'across the board'. As a result, a national accreditation control board was established.

1978: Work in several other countries led, in 1978, to the proposal that an ISO technical committee should be set up to produce a standard for quality assurance.

1979: BSI published BS 5750 (Parts 1, 2 and 3). These quality systems included some substantial improvements to BS 5179 and could be used as contractually enforceable in the commercial sector.

It should be noted, however, that it also retained some of the omissions and failings of BS 5179 and the defence systems.

1980: ISO/TC 176 'Quality Assurance' was established with UK (BSI) and Canada (SCC/CSA) providing the secretariat. The first meeting was held in May 1980 and the UK subsequently adopted the full secretarial role of the committee.

1981: BSI published BS 5750 (Parts 4, 5 and 6) guidance documents explaining the intent and practical interpretation of Parts 1, 2 and 3.

Parts 4, 5 and 6 became obsolete with the updated publication of BS 5750 (1987) but are expected (at the time of writing) to be re-issued in an updated form to provide guidance on the interpretation/application of BS 5750 (1987) Parts 1, 2 and 3.

1983: Following considerable internal reorganisation and discussion within the ISO/TC 176 committee, the structure and modus operandi was finalised, and in October 1983 committee ISO/TC176 held its first meeting having a UK secretariat.

The first draft for public comment of ISO 9001-4 appeared.

1984: The National Accreditation Council for Third Party Certification was established by the UK government. This body reports directly to the Secretary of State for the Department of Trade and Industry. Its objective is to ensure that all eligible certifying bodies meet certain published acceptance criteria and that they, in turn, are also assessed.

1985: ISO 8402 was produced, to replace BS 4778, with internationally recognised terminology for quality. LRQA started to offer quality system certification in the UK.

1987: In March, the ISO 9000-4 series of documents was made available abroad and, in June, became available in the UK as BS 5750 Part 0 (in two sections) and Parts 1 and 2. Part 0 is guidance to Parts 1, 2 and 3, which will be fully described in this chapter and amplified in Chapter 4. The relationship between the BSI and ISO documents is also explained in the following sections.

In September 1987 the first drafts of Parts 4, 5 and 6 were circulated privately for comment.

Calibration (BS 5781) has yet to be added to the ISO series.

The Committee for European Normalisation (CEN) adopted the ISO 9000 series as the quality management system and standard for use across the twelve European member states of the European Community and the European Free Trade Association. EN 29000-4 series was now the European national standard.

The European Community established criteria and recommendations for the mutual operation and recognition of accreditation and certification systems in its individual member states.

With the launch of the ISO 9000 series, LRQA started to offer quality system certification throughout Europe.

2.2 THE PRINCIPLES OF BS 5750

BS 5750 is the UK National Standard for quality systems. It tells suppliers and manufacturers what is required of a quality-orientated management system. It does not set out unusual or special requirements which only a few firms can or need comply with, but is a practical standard for quality systems which can be used by all.

The principles of BS 5750 are applicable whether a company employs 10 people or 10 000 people. It identifies the basic principles and specifies the procedures necessary to assure that the goods leaving the factory meet the customers' requirements.

It is important to understand that BS 5750 is not a product specification. It says nothing about product performance or even how to specify products. Its purpose is to ensure that the products are made under the control of a system which ensures that they do meet specification on time and at an economic cost. Thus BS 5750 and the product specification work together, and both should be specified as a requirement.

Whilst the principles of BS 5750 are widely applicable, there are various factors which will affect the way in which the requirements are implemented in different organisations. These result in quality systems of varying complexity in different situations. Some such factors are:

—extent of executive personal involvement
—workforce commitment and loyalty
—complexity of company structure
—physical considerations (multi-site, etc.)
—purchasers' quality requirements
—statutory obligations
—supplier involvement in production cycles

2.3 THE PARTS OF BS 5750 (1987)

BS 5750 recognises that the more complex and sophisticated products often require more extensive systems to control the production activities and hence to assure the quality. It is therefore effectively structured as three standards in one (Parts 1, 2 and 3), with the requirements running from the most comprehensive to the most simple in the same numerical order (see Table 2.1).

These three parts are accompanied by Part 0, which is in two sections and provides guidance on the use and implementation of ISO 9001, 9002 and 9003. Part 4 provides more detailed guidance on the interpretation of the various sections of Parts 1, 2 or 3, in much the same way as does Chapter 4 of this book.

TABLE 2.1

BS 5750	ISO	Title
Part 0	—	
Section 1	9000	Guide to principal concepts and applications (guide to selection and use)
Section 2	9004	Guide to principal concepts and applications (quality management and quality systems elements—guidance)
Part 1	9001	Specification for design manufacture and installation
Part 2	9002	Specification for manufacture and installation
Part 3	9003	Specification for final inspection and test
Part 4	—	Guide to the use of BS 5750, Parts 1, 2 or 3

It is interesting to note that the principal parts of BS 5750 (1987) betray its ancestry in companies that closely followed the arrangement of the Allied Quality Assurance Publications (AQAPs).

In practice it has been found that certain products and industries have special or unusual practices or requirements which are not reflected in Parts 1, 2 or 3. In such cases a supplementary document (e.g. Quality Assessment Schedule) may be needed to link the product requirements to those of the system. These QASs are industry-specific interpretations of BS/ISO.

No principal part of BS 5750 (i.e. Parts 1, 2 or 3) is considered as being of a higher status than the others. Each part is appropriate to given circumstances and is selected accordingly. It is, however, true to say that the requirements of Part 3 are encompassed by Part 2 and likewise the requirements of Part 2 are encompassed by Part 1.

It is intended that the standard only (Parts 1, 2 or 3) should be called up in a contract; the guidance notes are to assist with interpretation and application of the standard, and are not mandatory requirements.

It is important to note that the standard defines *minimum* requirements, which are not necessarily comprehensive. Other supplementary requirements may be quoted in the purchaser's contract.

Chapter 4 will describe the purpose of each paragraph of BS 5750 Parts 1 and 2 (and hence ISO 9001 and 9002) in detail.

Some limitations of BS 5750 at the time of writing (1990) include:

—no servicing clauses in Part 2;
—no purchasing clauses in Part 3;
—there are no references to BS 5781 (Calibration).

TABLE 2.2
UK National Standards Cross-References

MOD quality topic	MOD 05-21 quality system	MOD 00-16 Software QA	AQAP 1 Edition 3 (replacing 05-21)	AQAP 13 (replacing 00-16)	BSI 5750 Parts 1–4
Management organisation	202		202	202	4.2, 4.8(a)
Terms of reference					
Project organisation					
Management and control					
Assignment	204	6.1	211, 205(b)	205(b)	4.2.1, 4.5
Exception reporting	205	6.1.2	205(d)		4.12, 4.6
Event reporting (project diary/logs)	205		205(d)		4.6
Specific project standards	208	6	205(a)	205(a), 208	
Development aids (support software	204	6.3, 6.4, 7.3, 7.8		211	
and tools across whole life-cycle)	213	6.9, 7.9, 7.9.1		217	
Handover to client	205		205(a)		4.6
Project filing system					
Planning	203				4.4
Estimating	203				
Scheduling activities	203	7.1	204, 211	207	4.2.1
Life-cycle planning		5.4	204	204, 211	4.4, 4.8(a)
Quality planning		5.5, 7.1, 7.8	204	207	4.4, 4.8(a)
Project planning		6.1.1, 7.1	204		
Design control	207		207	211	4.8
Requirement specifications		5.1	207, 208(a)	205(a)	4.5, 4.8(b)(e)
Functional specifications		5.3	207, 208(a)	205(a)	4.5
Review procedures	207	7.1	206	206, 207	4.3

Design		6.7, 7.2, 7.5.2, 7.7	207	207	4.8(j)
Product		7.5.7		207	4.8(j)
End phase		6.7, 7.5.3		207	4.8(j)
System					
Start-up and close-down	207		207		
Restart and recovery	207		207		
Application—security	207	6.4	207		
Design calculations (performance modelling)	207		207		4.8
Testability		6.4, 7.2		213	4.8(b)(c)
Design aids				205(a), 211	
Modelling	204	6.5		205(a)	
Prototyping	204	6.3, 7.3		205(b)	
Standard design forms	204, 207, 208	6.5		205(a)	
Programming activities	204	6.2, 7.2	205(c)	205(a), 211	
Programming specifications		6.3, 7.3	207, 208(a)	205(a)	4.8(e)
Software design techniques	204	7.3		205(a)	4.8(c)
Coding conventions				205(a)	
Programming standards				208	
Configuration control (build state)		6.4, 6.6, 6.6.4, 7.6, 7.6.2, 7.8			
System design					
System interfaces	207	6.5	207		4.8(d)
System elements	207	6.5	207		
Performance objectives	207	6.5	207		
Resource (memory) budget allocation		6.5			
Testing and inspection	213			213	
Test plans	205, 206, 211, 213	6.8, 7.3, 7.5.3, 7.8	205(c), 211	211, 213	4.12
Systems test group and terms of reference	204, 205, 213	6.8		213	4.6, 4.7, 4.14

(continued)

TABLE 2.2—continued

MOD quality topic	MOD 05-21 quality system	MOD 00-16 Software QA	AQAP 1 Edition 3 (replacing 05-21)	AQAP 13 (replacing 00-16)	BSI 5750 Parts 1-4
System test dossier	205, 208	6.5	205(d), 208(a)	213	4.6
Test data					
Test files					
Test results		7.8			
Systems test specifications	208, 213	6.5, 6.8, 7.5.3, 7.8	211	205(a), 213	
Systems test tasks	204	6.8	205(c)	213	
Systems test control		6.8	205(c)	213	
Systems test discrepancy reporting		6.6.1, 7.8	206	206, 213	
Systems test execution	204, 205, 206, 208, 213	6.8, 7.8	205(c)	213	4.6, 4.7
Systems test trial records		6.5, 6.8, 7.8	205(d), 208(a)	213	
Systems test development file		6.8, 7.8	205(d)	213	
Systems test review	207	7.8.2		207, 213	
Systems test techniques	209, 211			205(a), 211, 213	
Program test design	204, 211	6.3		211, 213	
Defect data collection and analysis—all levels		7.8	205(d), 206, 207	206	4.8(k)
Acceptance test procedures		6.9, 7.8.2, 7.9		211, 213	4.14
Documentation and change control	208	7.5			4.9
Document production	208	7.5.1			4.9
Amendments/change control	208	5.3.1, 7.6, 7.10	205(d), 208(b)		4.9(c)(d)
Documentation		6.5, 6.6.3, 7.6.3	208(b)		4.9(b)
Software		6.6.2		208	
Obsolete document removal	208		208(a)		4.9(e)

Software—library and configuration management		6.4, 6.6.2, 6.6.4, 6.10, 7.6, 7.6.2/4, 7.10		208, 211	4.17
Operational documentation		6.5, 6.10, 7.5.4	208(a)		
Maintenance documentation		6.5, 7.5.5, 7.9.1, 7.10	208(a)		
Discrepancy reporting/corrective action	206	6.6.1, 7.6.1, 7.10			4.7
Technical reports/design reviews	206	6.7, 7.7	206, 207	206, 207	4.7
Progress meeting	207				4.2.3
Progress review reports	203, 206	6.1.2, 7.1			4.7
System test problem reports	206	7.8	206	206, 213	4.7
Problem reporting		6.10	205(d), 206	206	4.7
Miscellaneous					
Sales proposals		5.2, 6.10	208(a)	210	4.8(d)
Subcontracting software		6.11, 7.5.6, 7.11	203	203	4.3
System reviews	201	7.6.5	201, 202	201, 202	4.1, 4.2.2
Overall quality system					4.10
Inspection, test and measuring equipment	209		209		4.11, 4.13
Purchasing	210, 212		210, 212		
Preparation for delivery of software				217	
Records retention					4.6
Software support service					
Purchaser's representative					4.2.3
Health and safety statutory requirements					4.8(g)
Control of non-conforming material					4.16
Feasibility study standards		5.1	207, 208(a)	205(a)	4.5, 4.8(b)(e)

TABLE 2.3
Quality standards in various countries

Country or body	Overall quality management	System for design, manufacture, installation and servicing	System for production installation	System for inspection and test	Guidance documents
ISO	ISO 9000	ISO 9001	ISO 9002	ISO 9003	ISO 9004
Europe (CEN)	EN 29000	EN 29001	EN 29002	EN 29003	EN 29004
UK (BS)	BS 5750 Part 0	BS 5750 Part 1	BS 5750 Part 2	BS 5750 Part 3	BS 5750 Part 0
NATO	—	AQAP 1	AQAP 4	AQAP 9	AQAP 2 & 5
UK MOD[a]	—	05-21	05-24	05-29	05-22 & 25
Ireland	IS 300 Part 0	IS 300 Part 1	IS 300 Part 2	IS 300 Part 3	IS 300 Part 0
USA	ANSI/ASQC Q90	ANSI/ASQC Q91	ANSI/ASQC Q92	ANSI/ASQC Q93	ANSI/ASQC Q94
USA defence		MIL-Q-8958A	MIL-I-45208A		—
Canada	CSA Z299, 0-86	CSA Z299, 1-85	CSA Z299, 2-85	CSA Z299, 4-85	CSA Q420-87
Belgium	NBN X 50, 50-002-1	NBN X 50, 50-003	NBN X 50, 50-004	NBN X 50, 50-005	NBN X 50, 50-002-2
Denmark	DS/EN 29000	DS/EN 29001	DS/EN 29002	DS/EN 29003	DS/EN 29004
France	NF X 50-121	NF X 50-131	NF X 50-1322	NF X 50-133	NF X 50-122
Italy	UNI/EN 29000	UNI/EN 29001	UNI/EN 29002	UNI/EN 29003	UNI/EN 29004
The Netherlands	NEN-ISO 9000	NEN-ISO 9001	NEN-ISO 9002	NEN-ISO 9003	NEN-ISO 9004
Norway	NS-EN 29000	NS-EN 29001	NS-ISO 9002	NS-ISO 9003	—
West Germany	DIN ISO 9000	DIN ISO 9001	DIN ISO 9002	DIN ISO 9003	DIN ISO 9004
USSR	—	40.9001-88	40.9002-88	—	—

[a] This is now superseded by the AQAP and ISO standards.

2.4 COMPARISON OF STANDARDS—UK

For purposes of comparison a table of national standards (Table 2.2) is provided. It provides, by subject area, the cross-references of AQAP (Allied Quality Assurance Procedures), MOD (UK Ministry of Defence) and BS/ISO (British Standards/International Standards Organisation) paragraphs. Note that MOD 00-16 and AQAP 13 are included. These are software standards.

2.5 COMPARISON OF STANDARDS—INTERNATIONAL

Table 2.3 shows the equivalent standards for a number of countries.

Chapter 3

What Type of System?

3.1 A HIERARCHICAL ARRANGEMENT

Chapter 1 introduced the need for a quality management system and Chapter 2 described the currently available standards which provide guidance on the activities and controls that are required.

We must now tackle the question of how such a system might be structured. A typical and effective approach is to develop a hierarchy of procedures so that the general principles and rules are described in a top level document (the quality manual) which refers to subsequent levels of procedures and work instructions that provide the detail. These are a working set of instructions for day-to-day use within the organisation.

A quality manual describes the general quality policy and the organisation of the company, together with the responsibilities for quality. It then outlines the specific areas for control and references the lower level documents needed to carry out quality control and assurance.

For a particular organisation, some quality requirements might be described by a few simple statements. The manual itself then provides sufficient procedures and no lower level documents will be required. There are examples of two quality manuals in Part 4.

In the case of example manual QMS 001, in Part 4, management review and internal audit are catered for in the manual and do not necessarily require additional procedures at the lower level. The quality manual will provide a management overview and a top level reference to the areas of control.

Procedures describe:

—who does what
—what must be recorded
—how activities are planned

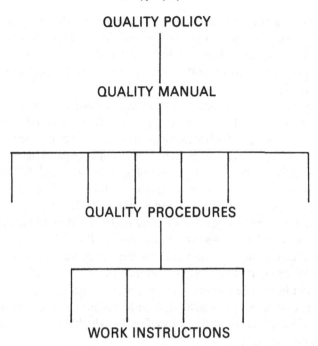

Fig. 3.1. Hierarchical relationship.

Work instructions provide detailed descriptions of activities and objective work standards which can be used as quality criteria.

Quality procedures and work instructions are referenced by the quality manual. These provide detailed instructions concerning routing of forms (e.g. defect sheets, purchase orders, concession notes, change notes, etc.), information to be recorded, responsibilities for each action, and so on. The document numbers of the procedures in the examples are described in Chapter 4.

Work instructions are a lower level of detail than the procedures. Whereas procedures describe how things are done and who does them, work instructions give specific information about standards to be achieved.

Figure 3.1 illustrates the hierarchical relationship of these documents.

3.2 THE DESIGN AND MANUFACTURING CYCLE

The hierarchy of quality system documents is complementary to the design and manufacturing life-cycle.

The idea of a life-cycle provides a convenient model that serves two purposes. Firstly, it allows one to represent the process of design and production in a graphical and logical form and, secondly, it provides a framework around which quality control and assurance activities can be built in a disciplined manner.

The conception, design and use of either hardware or software is an evolutionary process. That is to say, it is produced through successive iterative stages of specification, design and modification. Each stage of evaluation and, later, tests and field use results in changes. This evaluation process should involve successive tiers of specification and design where each step is verified against the requirements of the preceding stage. Thus a form of reliability growth applies and a viable product is evolved.

This top-down iterative succession of activities is commonly called a life-cycle, and is described in diagrammatic form in Fig. 3.2.

The first stage is the definition of requirements. Always the first stage in any problem-orientated process, it is, in fact, the most difficult to achieve. The problem lies in the communication between customer and developer. The former often does not know what he wants and, as a result, the latter will have difficulty in formalising a specification which accurately translates requirements into design.

Assuming for the moment a satisfactory set of requirements, the next stage is usually for the developer to respond with a functional specification. This equally important stage defines what the developer is to implement and thus provides the first level of visibility to the customer of the eventual product.

From here onwards there is design iteration of some form. At the highest level a system design is established which will allow the separation of software components from non-software components and the definition of the interface between them. Very often a hierarchical design will be generated in order to establish a framework. Software design is then performed by use of an established 'top-down' methodology. The lowest level in the case of software design provides the basis for coding and largely defines the structure of the program. The next stage involves testing at various levels.

In the case of simple hardware products, the functional specification may be followed simply by drawings. Firstly, the design itself is tested (qualified) against the specification and then, during manufacture, conformance to that design is measured by inspection and test procedures.

Finally, acceptance of the product will take place, at which stage the customer usually 'signs-off' the product, possibly with some defects still

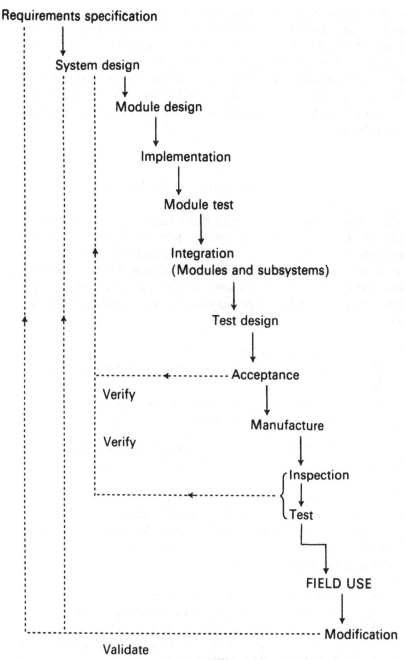

Fig. 3.2. Life-cycle diagram.

acknowledged. Acceptance may apply to the design or to the ongoing manufactured items. In-service or field use can also be viewed as part of the system life-cycle right up to the point of obsolescence.

It can be seen that where design is involved it is that, rather than the manufacturing part of the life-cycle, which is of paramount importance. Thus the requirements of BS 5750 Part 1 (ISO 9001), which are additional to those of Part 2 (ISO 9002), are those which impose standards and review on to the design cycle.

3.3 THE QUALITY PLAN

Whereas the quality system is generic and exists to control all design and manufacturing activities, the quality plan is product-, project- or development-specific. It should be produced for each and every development project. Whereas routine manufacture will be covered by the procedures and work instructions, a significant production contract, although not involving design, may call for the preparation of a quality plan:

—if the contract specifies it;
—if inspection and test activities, additional to normal routines, are involved;
—if customer involvement in inspection and test is specified;
—if different standards are to be used.

The quality plan should be produced as early as possible after the receipt of the contract. A draft plan may indeed have been a requirement as part of the tender.

The quality plan should follow a specified format which will be defined in the quality system. Either the quality manual or the procedures (depending on the quality system) will provide the guidance for its format and contents. Each stage of the design and manufacturing cycle must be considered.

The purpose of the plan is to set out:

—the standards and procedures to be used;
—the areas in which QA will be exercised and the means;
—the division of responsibility between QA and other functions;
—a formal record of any exclusions or waivers of procedures and standards which would otherwise apply;

—the means by which any specific QA standards called for in the contract will be met;

—the external and project QA reviews and audits which will be carried out;

—the audits which will be carried out;

—the design reviews to be held;

—the planned dates by which various QA activities must be carried out;

—the inspection and test strategy and procedures to be used;

—acceptance details.

Although format and layout will be common for a particular quality system, the following sections should normally appear in a project quality plan:

Introduction
—Description of the system.
—List of hardware and software deliverables.
—Company and QA organisation.
—Applicable references.
—Any specific contractual requirements for QA, test, acceptance, etc.
—Responsibilities for QA, inspection, test, etc.

Standards and procedures
—List of external and internal design or quality standards which are to be followed.
—List of internal design or manufacturing procedures which are to be used.
—Record of any exclusions or waivers which apply to the above.

Design cycle
—The stages of design shall be delineated by means of a life-cycle diagram. A generic example is shown in Fig. 3.2 of Chapter 3.
—A document hierarchy diagram will be provided covering all documents produced for the design. If necessary it will be amplified by a numbered document list.

Design reviews
—A list of design reviews will be provided such that each review can be related to a point on the design cycle.
—Each design review will be appropriately designated according to the stage at which it takes place (e.g. document review, code review, drawings review, test results review).

Inspection plan

—An inspection plan detailing each hardware inspection activity will be included.

—The detailed inspection procedures (e.g. goods-in PCBs, loaded PCBs, cable assemblies) will be referenced.

—Planned code walkthrough/inspections for software will be detailed by module. The methodology for the inspection will be described or referenced.

—Unless a waiver is given, code will be subject to static analysis. Details of the package to be used and the schedule of activities will be provided.

Test strategy

—A strategy for the sequence of functional testing will describe the levels of test to be applied (e.g. incoming hardware, coded modules with simulator, integration of items, final functional test, marginal, environmental and misuse tests).

—The relevant test procedures will be listed.

Subcontract

—By reference to the design cycle all items of hardware or software which are to be subcontracted will be listed.

—The QA arrangements for control of software and hardware designs will be described.

Acceptance

—Proposals for acceptance testing and demonstrations will be listed.

3.4 MANUFACTURING INSTRUCTIONS

The manufacturing instructions comprise the process instructions, inspection and test instructions, and the records that are generated.

They can overlap with the work instructions, which may or may not be one and the same. Depending on the size of the organisation, and the number of products in production, the number of distinct types of documents will vary.

There is no virtue in an over-complex structure of documents and the manufacturing and quality systems should be tailored to provide the minimum number of types and minimum quantity of procedures, consistent with an adequate control of the business.

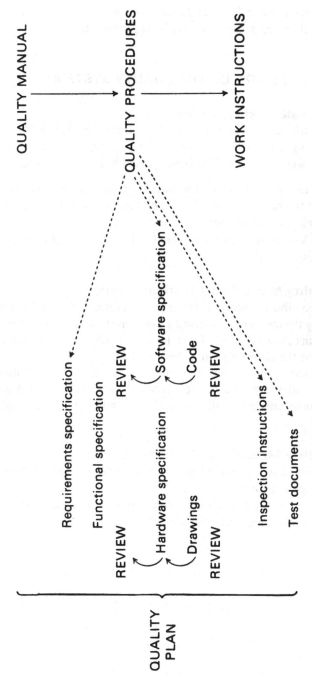

Fig. 3.3. Interrelationship of documents.

BS 5750 (and the other standards) does not seek to impose more procedures than are needed to meet its requirements.

3.5 SIZE OF THE QUALITY SYSTEM

3.5.1 A Single Quality Manual

Following the above statement it may transpire that a very small manufacturing organisation (say 2–5 people) could document its quality system in a single manual. This is not generally preferred since:

(a) It is inflexible at the level of inspection and test procedures, which need to change frequently. The manual would thus need to be re-issued for each change.

(b) It is less likely to give an overall policy view, being dominated by product detail.

3.5.2 Quality Manual plus Work Instructions

It may be possible to incorporate the procedures into the quality manual, leaving only the inspection criteria and various product-specific criteria to the work instructions level. Even in very small organisations this is preferable to the above option in Section 3.5.1.

Nevertheless, there is no particular advantage in combining the procedures with the manual just to increase the size of the latter. A number of small documents are easier to read, use and control than one large document.

3.5.3 Quality Manual plus Two Levels

The hierarchy shown in Fig. 3.1 offers the most flexible arrangement, even for small organisations.

Figure 3.3 shows the interrelationship of the various documents mentioned in this chapter to the life-cycle.

PART 2

A Structure of Manuals and Procedures

Chapter 4

Designing One's Own Quality System

4.0 INTRODUCTION

This chapter will discuss each of the major headings of a quality system for both development and manufacturing organisations. It has been arranged so that the following section numbers correspond with those of BS 5750 Part 1 and thus of ISO 9001, the latter having an identical layout and content.

The 'Part 1'/'9001' part of the series includes the quality systems requirements for the various activities and controls involved in design and development. Companies engaged only in manufacturing activities will require only the BS 5750 Part 2 (ISO 9002) systems, which are, however, a subset of the 'Part 1'/'9001' requirements and can, therefore, also be extracted from this chapter, as will be explained. This has been dealt with and the standards compared in Chapter 2.

The particular requirements, and the associated procedures in Part 4, which are needed only for 'Part 1'/'9001', are printed in italic, as are the procedures in Table 4.1. Those in normal print are the 'Part 2'/'9002' requirements which are common to both 'Part 1'/'9001' and 'Part 2'/'9002'.

This chapter adopts a top-down approach whereby the procedures are developed to meet the top level management requirements of BS/ISO. This does not imply that a company need impose tasks or procedures that are not necessary for the conduct of the business. The concept is that the quality systems and procedures should be an integral part of the activities of running the business in such a way as to benefit that business overall. It is a prerequisite of a quality system that it assists the conduct of the business and ultimately creates profit by saving more on defects than it costs in prevention.

TABLE 4.1

	Company A (hardware)	Company B (hardware and software)	Company C (software only)
QUALITY MANUALS (4.1/2)	QMS 001	QMS 001	QMC 001
QUALITY PROCEDURES			
Contract review (4.3)	QPS 001	QPS 001	QPC 001
Configuration and change control (4.5)	QPA 002	QPB 002	QPC 002
Design and review (4.4)	*QPS 003*	*QPS 003*	*N/A*
Software development (4.4)	*N/A*	*QPB 004*	*QPC 004*
Software design standards (4.4)	*N/A*	*QPB 004*	*QPC 005*
Software review (4.4)	*N/A*	*N/A*	*QPC 006*
Subcontract equipment and software (4.6)	N/A	QPS 007	QPS 007
Documentation format and standard (4.5)	QPA 008	QPB 008	QPC 008
Purchasing and suppliers (4.6/7)	QPS 009	QPS 009	QPC 009
Control of production (4.9)	QPA 010	QPB 010	Covered in QPCs 5, 15, 17
Goods inwards and inspection (4.10/11)	QPA 011	QPB 011	QPC 011
Identification and traceability (4.8)	QPS 012	QPS 012	Covered in QPCs 2, 5, 8
Stores (4.15)	QPS 013	QPS 013	QPC 011
Line inspection (4.10)	QPA 014	QPB 014	N/A
Test (4.10)	QPA 014	QPB 015	QPC 015
Calibration (4.11)	QPS 016	QPS 016	N/A
Non-conformance, corrective action and records (4.13/14)	QPA 017	QPB 017	QPC 017
Quality records (4.16)	QPS 018	QPS 018	QPC 018
Internal quality audit (4.17)	QPA 019	QPB 019	QPC 019
Education and training (4.18)	QPS 020	QPS 020	QPS 020
Servicing (4.19)	*QPS 021*	*QPS 021*	*QPC 021*
Statistical techniques (4.20)	QPS 022	QPS 022	QPC 022

Chapter 8 describes, in some detail, the balance between prevention, measurement and failure costs. There should be no question of imposing unnecessary procedures solely to meet external requirements, and it is no part of BS/ISO standards to reduce the efficiency of a business.

It is a common misconception that quality systems consist only of additional inspection and test. This is not the case since a realistic approach will balance the benefit of prevention activities against the savings which they generate.

The template procedures in Parts 4 and 5 are not particularly lengthy since it is recognised that, for small to medium-sized operations, the important activities can be described in a few pages. The user of this book is urged to review the procedures offered and to tailor them to the requirements of his operation.

The following sections of this chapter will describe the requirements of each part of BS 5750 Part 1 (ISO 9001) and will refer to the template quality manuals (Part 4) and procedures (Part 5) which might be tailored to different types of company.

It cannot be claimed that such an approach will create a system which automatically brings a company up to the required standard. It will, however, ensure that the management addresses each of the essential areas of control and at least attempts to produce a procedure for each. Appropriate consultancy effort can then be sensibly employed having used in-house efforts to produce the initial framework.

Organisations who wish to satisfy only the manufacturing quality requirements ('Part 2'/'9002') should utilise this book but ignore the sections and references that are printed in italic.

Table 4.1 is a list of the procedures in Parts 4 and 5, and these will be referred to throughout the rest of this chapter. It is an important key to the rest of this book. It provides a complete list of the manuals and procedures that are provided as templates. The rest of this chapter will refer you to these documents as it explains each of the headings of the quality system which you are constructing.

Note: If you require a BS 5750 Part 1 (ISO 9001) quality system in order to satisfy the requirements of design and development, then use all of the headings in the chapter. If you require only BS 5750 Part 2 (ISO 9002) since you will only be manufacturing, then *ignore the sections and documents in italic script.* Whereas Companies A and B could require Part 1 or Part 2, Company C can only talk in terms of Part 1 since its product (software) consists of design.

The paragraph numbers used in this chapter and in the table correspond to the BS 5750 Part 1 headings. The numbering of the Part 2 headings is slightly different due to the omission of the design and development items. The Part 2 equivalent heading number is indicated in the following paragraphs.

Some of the procedures in Part 5 refer to Work Instructions (e.g. Drafting standard). Although these are not provided in this book, Chapter 5 provides some guidance and a limited number of examples.

Three hypothetical companies are referred to, covering a wide range of hardware and software products:

Company A: a small electrical development and manufacturing company.

Company B: a small software systems development house.

Company C: a vendor of software packages who also offers support.

In Table 4.1 there are three conventions:

(1) QMC 001, QPA xxx, QPB xxx and QPC xxx: in these cases there is a separate procedure template provided for each of the company types.

(2) QMS 001 and QPS xxx: this header signifies that a single procedure is given for two or even all three of the companies.

(3) N/A: signifies 'not applicable' where the procedure heading in question is not needed for that company.

(∗) The two quality manuals are QMS 001 and QMC 001. Thus a single template is suggested for Companies A and B, whereas a separate manual is suggested for Company C.

In small organisations it is not essential that every requirement of BS 5750 involves a separate procedure. In some cases (as, for example, the organisation structure or even job descriptions) this can be dealt with in the quality manual.

The following notes should be remembered when using the templates:

—Companies A and B make use of the engineering and quality manager role whereas Company C gives the responsibility to the technical director. QPS 007 spans both B and C companies, and thus the title will need to be adjusted accordingly by the reader.

—Some procedures refer to specifics (e.g. the computer language SYNON/2). Since they are templates the reader will need to adjust them according to the company's specific business.

—In some of the QPCs forms are referred to but not actually given. Again these are left to the reader to design for the specific company.

Table 4.1 shows template procedures and against each heading the BS 5750 Part 1 1987 reference is given.

4.1 MANAGEMENT RESPONSIBILITY (4.1 in Part 2)

This first section of the BS 5750 Part 1 1987 quality system requirements calls for:

—a policy
—an organisation structure
—regular review of the system

The policy of a company is made visible by manuals, procedures and documented records of quality activities. The quality management system described in this book provides the basic statement of policy. The quality records provide the confirmation that it is being implemented.

A key requirement is that the responsibilities for carrying out the various activities are clearly written down. An organisation chart is essential and is very often given in the quality manual (e.g. QMS 001).

It is not sufficient merely to appoint a quality manager. The chain of responsibility must start at the top and follow a clear path to those who carry out the quality tasks. Job descriptions covering the quality function should exist either in the quality manual or as separate documents and there needs to be a formal requirement for training (QPS 020).

In defining the reporting structure and responsibilities it is important to ensure that measurement/test, audit and review are always 'driven' by someone other than those who carry out the work. This does not mean that a person cannot inspect his/her own work but that an independent arbiter is always called for.

In small companies the appropriate quality resources are not consistent with having a full-time quality manager. The combined function of engineering and quality manager is often found and need not cause a serious problem. Clearly, the engineering manager part of the role involves responsibility for the design whereas the quality manager role involves review and audit of the design and process control aspects of the quality management system.

The procedures should thus ensure that quality assurance activities which involve the design/engineering function establish the role of an auditor who is independent of the engineering and quality manager. In QMS 001 an engineering and quality manager role is used as a possible arrangement for Company A and for Company B. In QMC 001 the technical director carries the quality responsibility.

In slightly larger companies the quality control activities (basically inspection and test) are sometimes, but not always, separated from the quality assurance activities (essentially review, audit, design standards and manufacturing standards). Arguments abound as to whether or not the former should or should not report via manufacturing. In practice it matters little because the success of the quality management system depends almost entirely on the attitudes from the top; if the chief executive is seen to disregard or shortcut quality procedures, or to allow expediency to regularly compromise quality, then any system is doomed to failure. Conversely, if that same individual participates in meaningful reviews and is seen to believe in the system then success is not far away.

Superficially QMS 001 and QMC 001 appear similar, as they must, if they are to reflect the top level requirements of BS 5750. Sections 4.9 to 4.16, however, are markedly different.

There is frequently confusion between the meaning of review and that of audit. The latter involves checking that procedures are correctly implemented. Review, as called for in this section, means posing the question 'Is the system adequate for our needs?'. System reviews, which are quite separate from audits, should be scheduled by the quality manager and must involve even higher levels of management. Typically, for a small firm, the managing director and the engineering and quality manager would carry out the review. The reviews might be approximately twice yearly and should be planned well in advance and tackle a different aspect of the operation each time. This schedule should not inhibit more frequent reviews whenever a significant event occurs in any aspect of the organisation or its products.

4.2 QUALITY SYSTEM (4.2 in Part 2)

The quality system is described in Chapter 3 and will probably consist of three tiers of documents:

—Quality manual
—Quality procedures
—Work instructions

Quality plans, inspection and test documents, quality records (resulting from all these activities), standards, calibration documents and so on will be referenced by the three tiers of documents which will need to describe them and provide sample formats.

The following sections will point to specific sample procedures in the QMS, QMC, QPS, QPA, QPB and QPC examples in this book.

The top of the quality system hierarchy is the quality manual. In Part 4 two sample quality manuals are provided.

QMS 001: this quality manual is laid out, like this chapter, using the section breakdown of BS 5750 Part 1 (ISO 9001) 1987. It contains a sample organisation chart (QMS 001, Fig. 1) and lists a typical set of quality procedures, as listed in Table 4.1. QMS 001 is considered suitable for Company A and Company B as described earlier.

QMC 001: for Company C the quality manual is laid out in a similar manner but the needs of a software house are catered for.

4.3 CONTRACT REVIEW (4.3 in Part 2)

The primary concern of this section is that the quality system ensures that each of the conditions of contracts taken on can be met. Review by all relevant parties, *before commitment*, is essential. The important feature of the procedure should emphasise the need for requirements to be clearly understood by both parties to the contract. Agreement at all stages is vital and steps for this should be clearly specified. QPS 001 describes a system which Company A or Company B might adopt. QPC 001 provides a system for Company C.

4.4 DESIGN CONTROL (not in Part 2)

It is this area which describes the major difference between BS 5750 Part 1 and Part 2 since Part 2 is concerned only with manufacture. The need is for one or more procedures to control the way in which design is carried out so as to meet the requirements of the contract or the project in question. Features include:

Project planning
Reviews
Selection and use of standards
Documentation format
Configuration control (i.e. documents, changes, build states)
Organisation
Verification and design tests

A simple single procedure for design and its review is sufficient for Company A and this is QPS 003. For Company B the additional controls to take account of software are contained in QPB 004. Company C, on the other hand, requires a rather different set of procedures since its design activity is concerned only with software and its changes and their maintenance. QPC 004, 005 and 006 describe a typical quality system.

4.5 DOCUMENT CONTROL (4.4 in Part 2)

This concerns the format and nature of the document hierarchy, and the way in which issues are identified and changes are controlled. In the former case a procedure is necessary which describes the document hierarchy (requirements, functional specification and so on, down to the level of drawings, component lists and software source code listings) and the purpose and format of each document. QPA, QPB and QPC 008 provide examples for each of the three types of company.

It is also necessary to be able to identify a product (or design) at any moment in terms of its build state. That is to say, the documentation hierarchy with each document, at a given issue number, constitutes the description of a specific piece of hardware or software. In addition, changes which occur until it is 'frozen' into a subsequently defined build state (sometimes called a baseline) must be documented and controlled. QPA, QPB and QPC 002 address this for the three companies.

4.6 PURCHASING (4.5 in Part 2)

In order to ensure that items entering the company conform to the requirements, it is necessary to have good control of suppliers and subcontractors. This involves:

—assessing suppliers' and subcontractors' capabilities
—creating complete and unambiguous purchase orders
—verifying that the items conform to specification
—maintaining surveillance of suppliers

QPS 009 provides a template purchasing procedure for Companies A and B, and QPC 009 caters for the requirements of Company C.

QPS 007 deals specifically with subcontracted design or manufacture.

4.7 PURCHASER SUPPLIED EQUIPMENT (4.6 in Part 2)

The ever-increasing size and complexity of large projects has led to hierarchical contract structures involving chains of subcontractors. For this reason it frequently arises that the purchaser provides the supplier with items for incorporation into the equipment which he has contracted to design and/or manufacture.

It is necessary, therefore, for any quality system to allow for the control of these items so that:

- customer's items are incorporated;
- storage and maintenance are controlled;
- losses and damage are identified and dealt with.

QPS and QPC 009 address this aspect.

4.8 PRODUCT IDENTIFICATION AND TRACEABILITY
(4.7 in Part 2)

A crucial feature of quality control is that all hardware and software (from goods inwards to despatch and onwards) must be uniquely identified. This means that the build state must be clearly discernible in terms of the issue status of drawings, the traceability of component parts, and the inspection and test history of the item (be it hardware or software) in question.

This can be achieved in some cases by simple labelling of items. With batch production, involving large quantities of items (e.g. PCBs), a batch document may be suitable. In this case the items would still need to be individually serialised and recorded on the batch list, which would also record the build state, inspection and test status.

QPS 012 for Companies A and B, and QPCs 002, 005 and 008 for Company C and QPA and QPB 010 also address this area.

4.9 PROCESS CONTROL (4.8 in Part 2)

This area of quality management is concerned with the effective application of standard methods and criteria in all aspects of production and installation. The '3rd tier', referred to as work instructions in Chapter 3, is described, with examples, in Chapter 5.

The QPs for Companies A and B which apply here are QPA and QPB 010. Company C is covered by QPCs 005, 015 and 017.

4.10 INSPECTION AND TESTING (4.9 in Part 2)

The basis of quality control (rather than quality assurance) is the activity of measuring whether items (including software) conform to the requirements. Inspection and testing are the cornerstones of quality control and apply throughout the manufacturing cycle. Criteria should be objective and all results recorded.

A number of QP templates cover this area: QPA 011, QPB 011, QPC 011, QPA 014, QPB 014, QPB 015 and QPC 015.

4.11 INSPECTION, MEASURING AND TEST EQUIPMENT (4.10 in Part 2)

This refers to calibration. There are a number of essentials in this area, such as:

—traceability of measurement standards
—periodicity of calibration
—records
—storage, handling and segregation

QPS 016 provides a template for Companies A and B. It is unlikely that Company C will need a calibration procedure, but should it do so QPS 016 could well be adapted.

4.12 INSPECTION AND TEST STATUS (4.11 in Part 2)

The need to identify the status of items, in terms of what inspection and test activities have been carried out, is covered in QPS 012 and QPCs 002, 005 and 008.

4.13 CONTROL OF NON-CONFORMING PRODUCT (4.12 in Part 2)

The key requirement here is to ensure that non-conforming items (i.e. defects) are prevented from being used any further once they have been identified. The procedure must also provide the means whereby these defects are suitably sentenced for rework, scrap, return to supplier, concession to use, etc.

QPA, B and C017 provide three templates for consideration.

4.14 CORRECTIVE ACTION (4.13 in Part 2)

Closely related to Section 4.13, this requirement follows on to ensure that defects result in investigation in order to put in hand remedial action so that the particular cause is irradicated.

This may lead to changes involving:

—a process
—a procedure
—a standard
—a supplier

QPA, B and C017, which catered for Section 4.13, also cater for this requirement.

4.15 HANDLING, STORAGE, PACKAGING AND DELIVERY (4.14 in Part 2)

These areas are traditionally potential contributors of defects. Work instructions are needed to describe the specific requirements for safe and protective handling, storage, packaging and delivery.

This is perhaps best dealt with at the work instructions level (Chapter 5) for small companies and a separate QP is not included here. The requirement is mentioned in the quality manuals and, in fact, QPB015 contains specific work guidance on the handling of metal oxide semiconductors (MOS) for Company B.

4.16 QUALITY RECORDS (4.15 in Part 2)

It is only possible to implement remedial action and to ensure the application of standards and procedures if thorough quality records are maintained. QPS018 and QPC018 cover this area.

4.17 INTERNAL QUALITY AUDIT (4.16 in Part 2)

The quality manuals (QMS 001 and QMC 001) refer to the review of the quality management system whereby the fitness for purpose of the system is assessed. Audit, however, is the function of checking that the quality management system, as it stands, is being accurately applied. Its primary purpose is to reveal deficiencies in the way things are done compared with the written standards and procedures.

A secondary benefit, however, is to reveal if parts of the system are no longer appropriate, in which case a review is indicated.

The audits should be scheduled in advance and should attempt to cover the operation by means of regular samples. The results, and the evidence of remedial action, should all be documented. This is covered by QPA, B and C 019.

4.18 TRAINING (4.17 in Part 2)

Training is included in the quality management system requirements, since it is imperative that a formal system exists for ensuring that staff are aware of the quality standards and procedures. Training is important for personnel motivation (see QPS 020).

4.19 SERVICING (not in Part 2)

The quality management system does not cease to apply at delivery of a product. Frequently after-sales service is provided. This should be regarded as yet another process stage in need of standards and procedures, with the same controls. Note that service includes support and maintenance.

QPS 021 and QPC 021 describe this area.

4.20 STATISTICAL TECHNIQUES (4.18 in Part 2)

A frequent feature of quality control is the use of statistical sampling, which has many applications from defect trend analysis to assessing supplier capability. Since the incorrect use of sampling theory may lead to unacceptable quality levels, it is necessary for the quality management system to provide a standard for this activity.

QPS 022 and QPC 022 address this aspect.

Chapter 5

Other Working Documents

5.1 WORK INSTRUCTIONS—THE THIRD TIER

In Chapter 3 the hierarchy of documents, known as the quality system, was outlined. In general, a three-tier approach was advocated, whereby the quality manual provided a top level policy overview, with specific quality procedures at the next level showing how the policy should be implemented. Chapter 4 has outlined some typical systems consisting of quality manuals and procedures for the three companies, and Parts 4 and 5 of the book provide the templates referred to.

Below the procedures level there exists another type of document, referred to in Chapter 3: the work instructions. Whereas the quality procedures (QPA, QPS, etc.) describe *what is to be done*, the work instructions provide detailed guidance and standards for *how it is to be done*.

If only a small selection of template work instructions were to be provided here, then the book would be several times its present size, since the range of work practices which cover all types of hardware and software products is very great. This chapter, therefore, will describe some of the types of work instruction which are to be found and will give brief examples of their content. The user of this book will then need to decide which specific types of work instruction (and how many) are necessary in order to control the processes of the company in question.

Work instructions include:

—work standards
—quality plans
—inspection instructions
—test procedures

There is some overlap between work instructions and quality procedures, and there is no hard and fast rule for drawing a line between them. Indeed the option was suggested, in Chapter 3, of merging these two levels. QPB 014 (in-line inspection) is an example of such overlap and contains work standard information for the inspector.

5.2 WORK STANDARDS

These are documents that describe specific attributes or measurements of a product or operation, or provide detailed job activities. They provide a standard for the operator (or even the designer) and, at the same time, a datum for quality control. Typical work instructions might include:

—a design drafting standard;
—a soft soldering standard;
—packaging instructions;
—housekeeping instructions.

Extracts from these are provided at the end of this chapter.

5.3 QUALITY PLANS

These are called for in the higher level documents and increasingly in many contracts. The reason for a quality plan is to provide a product- or project-specific focal point which states the quality policy for that design or manufacture. It will give specific information as to the timing and types of reviews, inspection, test strategy and so on.

The following sections are typically found in a quality plan.

● *Introduction:*
    specific standards and references
    contractual quality requirements (e.g. witnessed tests)
    main deliverables
● *Organisation:*
    any project-specific quality responsibilities
    customer's quality representative
    project organisation

- *Documentation:*
    the document hierarchy for the product (including review and test procedures)
    timing of document reviews
- *Reviews:*
    type of reviews and timing
    methods (e.g. walkthroughs, static analysis, etc.)
- *Test:*
    overall strategy (e.g. top-down, bottom-up)
    list of test specifications
    acceptance test arrangements
    environmental tests, soaking, etc.
- *Records:*
    format
    specific customer records required
    certificates (e.g. conformity)
- *Subcontractors:*
    list of subcontractors and the surveillance arrangements
    list of bought-in equipment and software

For a contract involving design (BS 5750 Part 1) then probably all of the above headings will be required in the quality plan. For a manufacturing contract (BS 5750 Part 2) then the appropriate subset of headings can be chosen.

5.4 INSPECTION INSTRUCTIONS

These describe precisely what the inspector must measure in terms of:

—tools or test equipment to be used;
—sample sizes and pass/fail criteria;
—attributes or measurements to be judged;
—records to be kept.

In some cases the actual copy of the inspection instruction being used by the inspector may be used as the inspection and defect record. An inspection instruction need not necessarily be a large document. In many cases a few words (or diagrams) under each of the above headings will suffice. On the other hand, a large inspection task, involving many complex measurements and the use of sophisticated measuring equipment, may require a larger document.

5.5 TEST PROCEDURES

The philosophy here is much the same as for inspection instructions. Detailed step-by-step instructions are needed so that the exact sequence of tests, with their anticipated results, is standardised irrespective of the test operator.

Once again the copy of the test procedure may usefully serve as the test record and would therefore contain spaces for the tester to record results as the test proceeds.

5.6 EXTRACTS FROM WORK STANDARDS

5.6.1 Extract from a Design Drafting Standard

CONTENTS LIST

1. INTRODUCTION

2. DRAWING PRACTICE

3. DIMENSIONS AND LETTERING

4. DRAWING NUMBER SYSTEM

5. MICRO-FILMING

6. QUALITY ASSURANCE

7. METHOD FOR RECORDING VENDOR/CLIENT PRINT INFORMATION

8. DRAWING ISSUE

Figure 1 Modification Sheet Form

1 INTRODUCTION

This work instruction applies to those whose work involves drawings and associated documents.

1.1 Scope

This work instruction is concerned with systems from the production of drawings through to the presentation of drawings suitable for microfilming and suitable for paper copies to be obtained for use by a manufacturing department or by subcontractors.

Most techniques are an integral part of a draughtsman's skill and this document only covers practices which differ; it does not state the obvious.

Also included are item lists by the designers to enable manufacturing, or subcontractors to manufacture subassemblies, and to complete the final assembly.

All drawings and item lists shall be in line with this document.

1.2 References

This standard makes reference to:

BS 308: 1972	Engineering Drawing Practice
BS 3429: 1975	Sizes of Drawing Sheets
BS 3939	Graphical Symbols for Electrical Power Telecommunications and Electronics Diagrams
BS 5536: 1978	Specification for Preparation of Technical Drawings and Diagrams for Microfilming
BS 5370/6221	Printed Wiring (Design Manufacture and Repair)

2 DRAWING PRACTICE

2.1 Drawing Sheets

Drawing sheet sizes are based on the international 'A' series. These are pre-printed to a standard format on both paper and film.

There are five standard sizes of paper in use, as shown below:

ISO 'A' Series	*Size (mm)*
A0	841 × 1189
A1	594 × 841
A2	420 × 594
A3	297 × 420
A4	210 × 297

A drawing should be made on the smallest sheet size consistent with clarity and unambiguity.

A0–A2 drawing sheets should only be used when essential and are discouraged, since they lead to difficulty in storage, handling on the shop-floor and lack of clarity when reproducing on microfilm.

2.2 Pre-printed Drawing Sheets

The standard sizes A1–A4 drawing sheets are carried in stock.

3 DIMENSIONS AND LETTERING

3.1 Lettering

The individual style of a draughtsman is not restricted. All handwritten letters and figures must be in open form and devoid of serifs or other embellishments. All strokes should be black and of consistent density compatible with line work.

Capital letters are preferred to lower-case as they are less congested and are less likely to be misread when reduced in size. It is recommended that lower-case letters be restricted to instances where they form part of a standard symbol, code or abbreviation.

3.2 Character Height

The character heights shown in Table 1, whilst not mandatory, are recommended as good practice. It is stressed that these recommendations are for minimum sizes. However, when lower-case letters are used they should be proportioned so that the body height is approximately 0·6 times the capital letter height. The stroke thickness should be approximately 0·1 times the character height and the clear space between characters should be approximately 0·7 mm for 2·5-mm capitals; other sizes in proportion.

The space between lines of lettering should be not less than half the

TABLE 1
Drawing sheet sizes and character heights

Application	Drawing sheet size	Minimum character height (mm)
Drawing numbers	A0, A1, A2 and A3	7
	A4	5
Dimensions and notes	A0	3·5
	A1, A2, A3 and A4	2·5

character height but, in the case of titles, closer spacing may sometimes be unavoidable (refer to Table 1).

3.3 Presentation

All lines should be black, dense and bold. It is important that all the lines on a drawing, including those added in any revision, should be of constant density and reflectance. The lines on any one drawing sheet should preferably be entirely in pencil or entirely in ink. If a mixture of pencil and ink is used, every effort should be made to ensure that uniform density and reflectance are maintained.

Each type of line should be of consistent thickness throughout its length.

To allow for the limitation on print-out and viewing of reduced microfilm copies, adjacent lines should be spaced not less than 1 mm apart. It is accepted that in some cases the scale of drawings will be violated.

3.4 Dimensions

All dimensions on new drawings must be expressed in millimetres (mm) and decimals of a millimetre. Metres and centimetres should not be used. Imperial dimensions may only be used when related to the thread forms (US Standards).

The decimal sign is a point, which should be bold, given a full letter space and placed mid-height of the numerals.

When the dimensions are less than unity the decimal sign must be preceded by the cypher '0', e.g. 0·5 not ·5.

When there are more than four figures to the right or left of the decimal sign, a space should divide each group of three figures counting from the position of the decimal sign, e.g.

12 345·6
1 234·567 8
12·345 67

Dimensions should be expressed to the least of significant figures, e.g. 35 not 35·0. Each dimension necessary for the complete definition of the subject should be given on the drawing and should appear once only in any view. Each dimension should be specific and it should not be necessary for any dimensions to be deduced from other dimensions.

5.6.2 Extract from a Soft Soldering Standard

5 SOLDERING

A satisfactory solder joint shall have a good clean, shiny appearance (Fig. 5.1) without excessive or insufficient solder (Fig. 5.2) and free from defects, as detailed below.

(i) *Major defects*

Dry joints exhibit a distinct line where the solder joins the wire or high contact angle between solder and wire (Fig. 5.3).
Unsoldered joints (Fig. 5.4).
Bridging between joints and solder splashes (Fig. 5.5).
Blow holes; these may hide defective joints.

Minor defects

Pin holes shall not be visible to the naked eye (Fig. 5.6).
Moved joints; they have a frosty and granulated appearance (Fig. 5.7).
Solder spikes; ideally no solder spikes shall exist but if they do they shall not exceed 100% of the soldered conductor diameter and then only when no danger of shorting or damage to insulation is possible (Fig. 5.8).
Foreign matter shall not be apparent in a soldered joint.
Joint clearance shall be maximum 1·5 mm nominal unless otherwise stated (Fig. 5.9).

(ii) Tracks and lands shall not be lifted or damaged, caused by excessive heat during soldering (Fig. 5.10).

5.6.3 Extract from a Packaging Instruction

1 INTRODUCTION

1.1 Purpose

The purpose of this work instruction is to provide guidance for all those involved in packing and despatch of goods.

1.2 Scope

This work instruction is concerned with the method by which the product is prepared for shipment from the factory premises.

2 PACKING FOR SHIPMENT

2.1

Goods requiring despatch will be so indicated to the packing department by means of despatch notes in triplicate from the administration department (see QP XXX), together with any special relevant documents.

2.2

Depending on the type of goods, the packing department will determine the method of packing and despatch.

2.3

For small items (less than 5 kg) packing may involve special bags or bubble pack and boxes. At all times the items are to be protected against damage and packed accordingly.

2.4

For large items chemical foam may be used to provide full protection against damage within cartons of a suitable size. One or more items may be contained in one carton. If several items are packed together then care should be taken to separate them. For handling reasons, the guideline maximum weight for a carton is 40 kg.

2.5

Large items that cannot reasonably be despatched in cartons will be securely mounted on wooden pallets of appropriate size.

3 DESPATCH DOCUMENTS

3.1

Each item, or container, will be marked with destination and weight, and will be accompanied by the white copy of the despatch note after details have been entered on to the triplicated set. An adhesive clear envelope is provided for this purpose.

3.2

The Company's policy for export shipments is to freight forward through an agent specified by the customer. In these cases four copies of the appropriate invoice and packing list are to accompany the shipment to the agents for them to raise the necessary clearance documents.

4 ADMINISTRATION

4.1

The shipment weight is to be recorded on the pink copy of the despatch note and returned to the administration department. The yellow copy is to be returned to final inspection as proof of completion.

5.6.4 Extract from a Housekeeping Instruction

GENERAL HOUSEKEEPING

(1) Maintaining a suitable working environment is an integral part of the job description for all personnel.
   The following will apply:

   —Floors will be checked each morning and cleaned as necessary.
   —Shelving will be cleaned once per month by vacuum cleaner.
   —Partaking of food and drink will be away from the vicinity of equipment and components.
   —Partly or fully assembled equipment will be covered when not being worked on.
   —Smoking is not permitted in certain parts of the factory or offices as indicated.
   —Benches are the responsibility of individuals who will keep their workplaces clean and tidy.
   —Personnel will be instructed in the principles of good housekeeping.

(2) It is the quality manager's responsibility to ensure that these activities are carried out and he shall check them on a regular basis.
(3) The COSHH (Control of Substances Hazardous to Health) Regulations 1989 will be reviewed by the Engineering and Quality Manager and put into practice.

PART 3

Other Essential Guidance

Chapter 6

The Overview of Quality

6.1 THE THREE AREAS OF QUALITY

There are three quality levels of requirement for any company: quality economics at the company level, quality assurance at the project and product management level, and quality control at the day-to-day development level.

The concept of *quality economics*, at the company level, is usually appreciated and associated with business success whereas the distinction between quality assurance and quality control is often lost.

Quality assurance is part of management and organisationally it falls under the control activities of project management; it has to be implemented across all four project management activities (of planning, organising, monitoring and controlling) since all activities have to be controlled.

Quality control, on the other hand, is concerned with the day-to-day implementation of tools, methods, standards and procedures, and is essentially non-management.

These three aspects of quality are clearly identifiable and mutually dependent, one upon the other, as parts of an overall company quality programme; if any is missing or the links are weak, the whole quality system will become ineffectual and collapse.

6.2 SIX THINGS TO BE DONE

Every company has its own unique character. The means by which a quality programme can be implemented must therefore be tailored to the specific circumstances. Although the means can vary, the end results, namely improvements in quality and productivity, have to be the same.

Six points, crucial to the competitive survival of any company, are listed below. Ways of meeting these points will be looked at briefly in the next few paragraphs.

The six points, from W. E. Deming's list of 14 for industrial processes, are as follows:

(1) management commitment;
(2) implementation of methodologies;
(3) formalised product assurance, integrated with the methodologies;
(4) technological awareness;
(5) continual implementational improvement;
(6) employee participation.

6.2.1 Management Commitment

All managers, at all levels of the company, must lead the quality programme to improve quality and productivity. Passive support is not enough. There is a clear link between quality and productivity.

Management training must include quality education; there must be refresher training courses to keep abreast of commercial, competitive, product and technological changes.

Companies must prepare and implement a company-wide quality and productivity improvement plan, its implementation being driven from the top, and with individual managers being held accountable for quality and productivity in their areas.

Managers must lead and encourage their staff to ensure that their quality and productivity targets are achieved and maintained. Where they are not achieved, they must be held accountable. All managers need to understand (not appreciate) quality and the factors which influence it, since these factors are the determinants of price, delivery, competitiveness and ultimately of commercial survival.

Without exception management must state categorically that quality is everybody's responsibility and everybody must be actively encouraged to improve it.

Quality is the result of intentionality, not accident. It is a state of mind, founded on discipline and attitude, and manifested in an individual's overt behaviour. Quality is intangible: it does not appear in the statements of account or balance sheets; accountants and financial auditors do not recognise it. Without quality prudence, however, there cannot be financial prudence.

Management has to create a work environment in which quality and productivity can be implemented and the implementation measured, monitored and improved. This work environment must include both the macro (or company-wide) and micro (or project) levels. This environment must be structured and involve detailed planning and organising. This will lead to greater stability and control. Only with stability and control, at both macro and micro levels, is it possible to identify the causes of malfunctions.

The case for stability and, therefore, control is simple. Whereas a malfunction at the requirements stage costs 1 unit to correct, at the design stage the cost is likely to be 5 units. At software coding, for example, it rises to 10 units and during testing 20 units is more likely. By the time maintenance occurs it has reached about 30 units. These sums are still trivial when compared with non-conformance at the corporate level. A product brought late to the market, or a contract broken, can result in the collapse of a company.

Once stability has been achieved, circumstances become controllable; records can provide direct comparisons and accurate measurements can be made. Once identified, recorded and measured, areas of malfunction can be isolated and receive corrective attention.

The role of quality management is no less onerous than the rest of any other company management. Quality management must coordinate the quality effort, to communicate ideas and information on all aspects of quality and to monitor quality implementation.

Quality cannot be superimposed on a company or its projects by quality management. It must stem from the way in which the company operates, and attention to quality must be one of the company's principles.

In many respects quality is no more than the application of common sense and planning; there is no magic. Neither are there any 'at a stroke' solutions. None will be offered; there is only hard graft.

An important part of quality management is not to seek short-term cost saving, by making 'popular' decisions. There are grave dangers in making short-term assessments; the longer-term consequences must be considered.

6.2.2 Implementation of Methodologies

Next there must be the creation or acquisition and implementation of a formally defined life-cycle system. As a subset of the life-cycle methodology, any development that varies and is, therefore, unpredictable from project to project is nearly impossible to control; without control there cannot be improvements in quality and productivity.

A common life-cycle system— together with software and development, and production standards and procedures—promotes the re-use of requirements, design and code. A formal methodology provides many baselines from which lessons learned from experiences can be derived. From the records that comprise the baselines, the fine 'tuning' of quality, productivity and costing is possible.

6.2.3 Formalised Product Assurance, Integrated with the Methodology

To the Americans and the Japanese computing and software is about 'products', which are 'tangible'. Quality is about intangibles. Since it is easier to measure tangibles than intangibles, quality should be made tangible. We should try to concentrate on the tangible, hence the word 'product' can be used.

A formal product assurance organisation provides a mechanism for achieving an integrated life-cycle system, within the method for both hardware and software.

As much as 50% of a project's costs can be spent on repetitive and wasteful rework and retest; much development work is done badly and mistakes are repeated. As is well known, the cost of correcting early errors increases dramatically throughout the life-cycle. Early error detection, therefore, is the crux of improved quality and life-cycle productivity.

6.2.4 Technological Awareness

Companies must be keenly aware of advances in technology. They must forge links with major research initiatives (in the UK, for software, this would include Alvey and ESPRIT) and government programmes.

Internal quality evaluation programmes must also be aggressively followed. These programmes identify new policies, strategies, methods, tools and techniques for improving quality and productivity.

Not every good idea becomes good practice; there must be in-house evaluation and customisation if innovations are to be capitalised.

All these activities, however, require real management commitment and corporate backing and investment.

6.2.5 Continual Implementational Improvement

Once defined, a stable and controlled life-cycle and development process should be updated periodically to incorporate identified technological improvements (e.g. new methods, tools and techniques).

The introduction of any new technology must be carefully planned. Its role in the manufacturing system, software and development life-cycle must be clearly defined. Improvements in the life-cycle need to go through three stages:

(i) Measurement to establish a baseline and to define leverage points at which improvements can be made. A baseline cannot be established without records, and without records we cannot measure.

(ii) Innovation and evaluation to identify specific methods and tools to implement the improvements.

(iii) Transference of proven new methods and tools to managements and developers by training, education, standards and procedures.

As it is difficult to isolate the effects of individual new methods and tools any evaluation of their effectiveness must rely on overall trends.

The innovation and evaluation process must be carried out by macro and micro management, engineers, product assurance and measurement personnel.

6.2.6 Employee Participation

Any programme of quality and productivity improvement can succeed only with the enthusiastic support of employees. This is particularly true in the people-intensive computing industry.

To be successful any new method must be willingly embraced by its intended users. Moreover, employees are often in the best position to identify potential improvements. Methods of encouraging employee participation include quality circles, suggestion plans, technical training and continuing education. These create interest, involvement, inventiveness, discipline and the correct attitude.

From these six points one of the conclusions that can be drawn is startlingly simple. First time right production (or as it is sometimes called 'zero defects') is a realistic approach. To succeed, however, we need the right planning and organisation and it must be possible to implement suitable quantitative techniques over the tangible products.

6.3 ZERO DEFECTS

The implementation of the above six points is an approach to improving system, manufacturing software, and development quality and productivity. However, no single new method, tool or technique can overcome all the obstacles faced by software developers.

Organisations looking for a method or tool to produce order-of-magnitude improvements will be disappointed. Success can only come from a systematic approach using integrated systems. A manufacturing system, software and development life-cycle methodology is required, with the whole thing being management-driven.

The acquisition of new methods, tools, etc., is only one step towards improving quality and productivity. Organisations usually compound their difficulties by introducing new methods and tools into projects. The system, manufacturing software and development life-cycle process itself, not just individual projects, must be explicitly and overtly managed.

This approach leads naturally to a factory-like company organisation. Some of the key features of Japanese hardware and software factories include:

—Use of a detailed, precise and consistent life-cycle methodology, supported by equally detailed and precise standards and procedures.
—Great emphasis on quality management, using quantification techniques.
—Integrated and, where possible, automated micro and macro support systems.

However, to formally implement a quality and productivity improvement programme based on the 'six points', the company has first to be organised to support them. Without the organisation there can be no implementation of the points and no marked improvement in quality and productivity.

Although quality has been described as intangible, it is possible to quantify software to the nth degree, by measurement against standards of performance requirements and their non-conformance.

The Crosby approach (now also in wide use in Japan) described in *Quality Without Tears*, McGraw-Hill, Maidenhead, UK, 1984, attempts to make quality tangible. A quality product is one which meets all its performance requirements; only a quality product meets all its performance requirements. For this, however, requirements must be fully specified at the outset. There are, however, many performance requirements.

The performance requirements which have to be satisfied include those of the

(a) user, for functionality, reliability and usability;
(b) project manager, for systematic life-cycle, manufacturing system, software and development;

(c) quality manager, for production of a quality product;

(d) data administrator, for integrity and security;

(e) operational manager, for performance and machine resource usage;

(f) business manager, for delivery within schedule;

(g) contract manager, for lack of litigation;

(h) sales and marketing manager, for a satisfied client and 'good' press reports;

(i) shareholders, whether public or private, whose money is invested in the company.

The description of quality here is through the performance requirements, which are largely outside the development team. The quality programme then becomes outward looking, and not inward, as with many developments. The quality programme starts to become company-wide.

From this perspective, however, it must never be forgotten that a company must 'perform' at the macro level, and integrate all its internal and external systems.

Each of the performance requirements must have a detailed standard, comprising the many factors, steps, activities and functions which are to be satisfied, to conform to the performance requirement.

Quality measurement is then the sum total of all the factors, steps, activities and functions which achieve the performance requirements. The lack of quality is measured in terms of the cost of not meeting requirements. The costs of achieving quality are soon seen to be overtaken by the savings associated with non-conformance.

Zero-defect performance becomes achieveable if the requirements standards are clearly specified and preventive systems are in place. Record keeping is absolutely vital for measurement; only with records can quality and productivity targets and fine-tuning be established, on the road to zero defects.

As well as aiding the move to zero defects, measurement also makes visible the lack of quality and its cost.

Chapter 7

Auditing the System

7.1 WHAT IS AN AUDIT?

BS 5750 Parts 1 and 2 both call for a periodic and systematic audit of the quality system. In other words, the quality system must be regularly reviewed in order to check its fitness for purpose. Products, processes and working practices change and, as a result, the quality system will need to evolve in order to be effective.

A change in work practice or the addition of a new process may well entail that the procedures become outdated and that revisions become necessary.

Most managers are familiar with financial auditing and accept it as a means of monitoring the financial health of the company. Surprisingly, however, few understand and accept the need for a similar approach to monitoring the adequacy of work practices. The lack of internal audit and management review is one of the most frequent deficiencies, even within companies that have well controlled operations.

The distinction between review of the system and an audit of conformance to procedures has already been made in earlier chapters. It is important enough to repeat. We are concerned here with an open-ended review of the quality system to ascertain if it is adequate to assure the design and production of the goods or services in question. An audit is purely a matter of checking that people are operating to the procedures. Both activities have a part to play but the distinction should be clear. Audits of conformance to the procedures can be carried out at a slightly lower level than reviews of the system, which must involve top management.

Although similar, audit and assessment (the latter usually by an external person) are slightly different. Auditors (internal) are usually much more familiar with the company organisation and procedures than are assessors (external). Therefore auditors can examine the procedures in much more detail and be more constructively critical. Assessors, on the other hand, will usually need to exercise judgement in determining the acceptability of a company since time-scales will be limited. It is not expected, in the case of assessments, that every deficiency will be uncovered, only that an overall picture is obtained.

7.2 THE AUDIT PROCEDURE

In order to establish an audit and review system, the objectives must first be defined. Programmes and plans can then be developed and levels of responsibility and delegation laid down. The procedure needs to describe:

—How the audit will be carried out, and by whom.
—What features of the system will be examined.
—The time and place for the audit.
—How the audit will be reported and to whom.
—How corrective actions are to be implemented.
—What follow-up audits are to be carried out and how they will check the corrective actions.
—The procedure whereby deficiencies are agreed and recorded at the time.

It is important to ensure that:

—Actual practices are checked against the need to design and produce a product. This can be compared against specific standards and requirements as appropriate.
—Auditors are independent of the function being reviewed.
—Schedules are laid down in advance and that future audits are planned on the basis of the experience gained from current ones. In other words, problem areas, once revealed, should receive greater attention in future audits.

7.3 SYSTEM AUDITS

As described above, this concerns the adequacy of the procedures. It does not concern:

—the product and its design;
—conformance to the procedures.

The former is a product audit (see Section 7.5) and the latter is a procedure audit (Section 7.4) which is covered by the template procedures.

The subject matter for the audit is thus:

—the quality manual and procedures;
—actual working practices.

The references for consideration are:

● National and customer standards (these may have changed since the implementation of the quality system).
● Contract conditions.
● More recent guidelines (BS, Health and Safety Executive, etc.).

These audits are required, by BS and MOD, to be carried out at least yearly, as is explained in Chapter 9, which deals with third party assessment and accreditation. It is expected that the audit will have been carried out internally and the results will then be reviewed by the assessor.

The system audit should:

—firstly, review the defined procedures (i.e. the quality system) against the external requirements (e.g. BS 5750 Part 1 or 2);
—secondly, review the adequacy of the quality system in its implementation.

The former involves a paper comparison of the quality system with the standard. The latter requires the examination of the quality records themselves in order to search for problem areas.

7.4 PROCEDURE AUDITS

These are covered in the template procedures by QPA, B and C 019. They do not essentially address the adequacy of the system but check that the procedures are being implemented. Quality records (e.g. defect sheets, change documents, test records, routing cards) are sampled for completeness, accuracy and general conformance to the procedures concerning their use, circulation, content, storage and so on.

Procedure audits might well be delegated to a quality engineer although the results should be reported at quality manager level. Again remedial action plans must follow each audit, otherwise the exercise is pointless.

Procedure audits may reveal problems which require a change to the system. These should be brought to the attention of top management and result in a system audit (Section 7.3).

7.5 PRODUCT AUDIT

Product audits are independent reviews of the product quality and technical performance. They should not be confused with testing, which controls conformance to specifications and drawings as they stand. It is broader based and therefore permits the possibility of changes and enhancements to the test procedures.

The initial reference for a product audit is the functional specification. In addition, one would consult:

—catalogues and specifications
—manufacturing instructions
—customer contract requirements
—independent standards such as BS product standards
—legal requirements such as safety codes
—guidance documents such as the Health and Safety Executive document on the safe use of programmable electronic systems

The product audit will involve an examination of the product as a functional whole as well as a review of its component parts. Defect records will be examined for patterns of failure and problems of interworking of components and subassemblies.

In the case of software, the audit will effectively be a design review process where the ability of the code to meet the requirements is sought.

As was the case with the system audit (Section 7.3), special attention must be given to all factors which might affect product quality. Feedback from personnel on known problem areas is invaluable. The following are all contributing factors to product quality:

—packaging
—storage environment
—transportation
—installation and commissioning
—servicing and maintenance
—manuals and instructions

7.6 REPORTING

During an audit any observations should be recorded on to the audit checklists. These notes should be compiled into a formal audit report with a view to:

—providing a formal record
—assigning responsibility for remedial action
—agreeing the deficiency at the time
—making proposals
—reporting to top management
—providing a permanent record
—providing a checklist for future audits

In addition to the report, a management review is needed in order to:

—ensure that policies are being implemented
—reveal deficiencies at a high enough level to effect change
—discuss improvements
—verify that remedial action takes place

If a company is applying the audit system, a number of modifications to procedures will be generated over a period. This is a useful benchmark for external assessors since there should be evidence of changes. Audit frequencies should also be seen to change, as evidence that audit is being used as a selective tool in monitoring the needs of the business.

7.7 AUDIT SKILLS

There are no specific qualifications for carrying out an audit, thus anyone should be able to participate given that a comprehensive checklist is available. Nevertheless, the following characteristics are important:

—persistence
—tact
—thoroughness
—technical competence
—logical questioning
—familiarity with the standards in question

Independence is essential and thus it is common for managers to perform audits on other departments. The quality department is no exception and should also be audited no less frequently than any other.

Chapter 8

Collecting Quality Costs

8.1 TYPES OF QUALITY-RELATED COSTS

The need to identify the costs of quality is by no means new, but the practice is far from widespread. Attempts to budget for the various types of quality-related costs are rare and planning the activities to identify, measure and control them is even rarer. This is surprising in view of their relative size compared with profits.

Every company needs to establish a quality cost system which will collect costs across the whole life-cycle of the product, project or service to include user and operational failure costs. Only this approach will provide the total cost of quality across all the activities of a company and its clients. These costs can easily exceed 10% of sales and thus, by lack of control, can absorb the profit margin.

Quality costs can be categorised under two headings, i.e. operating quality costs and external assurance quality costs. These types and their subtypes are summarised in a recent international standard (ISO 1987). One interpretation is as follows:

Operating quality costs are those costs incurred by a business in order to attain and ensure specified quality levels. They include the following:
  (a) Prevention: costs of efforts to prevent failures.
  (b) Appraisal: costs of testing, inspection and examination to assess whether specified quality is being maintained.
  (c) Failure: costs (or losses) resulting from a product or service failing to meet the requirements (e.g. product service, warranties and returns, direct costs and allowances, product recall costs, liability costs).

External quality costs are described as follows:
External quality costs are those costs relating to the demonstration and proof required, by customers, as objective evidence. This includes particular and additional quality assurance provisions, data, demonstration tests and assessments (e.g. the cost of testing for specific safety characteristics by recognised independent testing bodies).

These groups will be described as part of a quality cost system of a manufacturing organisation.

8.1.1 Operating Quality Costs

Table 8.1 shows a generic breakdown of operating quality costs for a six-month period in a manufacturing and assembly organisation.

The total quality cost is £513 600 and sales for the six-month period are £6 000 000. The ratio of quality costs to sales is the usual way of expressing the relationship as a percentage of sales. In this case it is 8·57%. It is, however, well known that the reported costs of quality are usually lower than the actual costs and it is not uncommon for this cost of quality, as a percentage of sales, to be as high as 25%.

Returning to Table 8.1, the following activities should be performed as an attempt to minimise quality costs and it is important to record them.

Prevention
Design reviews. Reviews of drawings, specifications and test results at various stages in the design of a product prior to the release of drawings to production.

Quality and reliability training. A large company may justify a quality department, which includes staff with an understanding of quality. Small organisations must rely on a combination of external courses/consultants and on-the-job training provided by its senior staff.

Vendor quality planning. All vendors must be able to satisfy product or service requirements. If they do not, the company will produce inadequate products and both it and its vendors will suffer. A vendor's ability to meet requirements can be evaluated by assessments, surveillance reports, meetings with other clients of that vendor, questionnaires, product and company appraisals and audits.

Audits. Audits can be internal to the company, carried out by its own personnel or by independent auditors. Audits can also be external, on the company's vendors.

Installation prevention activities. They include a wide range of activities if

TABLE 8.1
Quality cost analysis report 1 January 1989–30 June 1989
(sales £6 000 000)

	Cost (*in thousands of £*)	Percentage of sales
Prevention costs		
Design review	1·5	
Quality and reliability training	6	
Vendor quality planning	6·3	
Audits	7·2	
Installation prevention activities	11·4	
Product qualification	10·5	
Quality engineering	11·4	
TOTAL	54·3	0·91
Appraisal costs		
Test and inspection	135·9	
Maintenance and calibration	6	
Test equipment depreciation	30·3	
Line quality engineering	10·8	
Installation testing	15	
TOTAL	198·0	3·3
Failure costs		
Design changes	54	
Vendor rejects	4·5	
Rework	60	
Scrap and material renovation	18·9	
Warranty	30·9	
Commissioning failures	15	
Fault finding test	78	
TOTAL	261·3	4·36
Total quality cost	513·6	8·57

contract conditions are to be satisfied. Examples are availability of proper tooling equipment and instrumentation, manuals, drawings, other documents and data. All these activities need to be planned, reviewed and audited.

Product qualification. This is the testing of a developing product against its engineering specifications, before the product's drawings are released to manufacturing.

Quality engineering. The preparation of quality manuals and quality plans relating to a particular product or service. Quality standards, procedures and work instructions are also included, as are reliability prediction exercises and the modifications which arise from their results.

Appraisal

Test and inspection. During production the product must be regularly tested and inspected against production specifications and quality engineering documentation. This will be done either by manufacturing or by quality personnel. They have different reporting lines and involve different overhead rates. Rework and idle time are excluded since they are both aspects of failure, *not* of measurement.

Maintenance and calibration. These are the costs of internal and subcontracted labour, and items needed to ensure the correct calibration, upkeep, usability, availability, and repair of all test and inspection equipment.

Test equipment depreciation. Test and measuring equipment will depreciate over accounting periods. It will also 'age' as technology advances. These costs should be easily obtained from existing accounts.

Line quality engineering. As well as preparing test and inspection documentation, many queries concerning their implementation will arise. This is a time-consuming process and needs to be properly reported and costed.

Installation testing. The installation and commissioning of products in many high technology environments requires careful planning by suitably qualified personnel.

Failure

Design changes. Defects revealed in manufacturing (or indeed later) may result in design changes. Some, or even all, of the activities detailed under prevention will have to be re-addressed as the defect is traced through the engineering and production cycle. In fact traceability back to the user specification and even to the contract might be needed. In the worst case, this may go right back to the marketing quality conditions.

Vendor rejects. Purchased items which are found to be defective must be reworked, accommodated, reclaimed from the vendor or written off. Procedures and standards for purchasing, receiving, storing and handling vendors' items and services are essential to minimise failure costs.

Rework. With each rework or work around, production costs are incurred. Design changes might also result from work arounds. With all rework, testing will be needed. In addition, rework will often result in idle time on other parts of the shop-floor, which must be costed.

Scrap and material renovation. This is the difference between the cost of purchased items which are found to be defective and any scrap or reclaim value obtained from the vendor.

Warranty. Any products recalled under warranty must be carefully investigated. Losses associated with a poor product or service and a falling reputation are difficult to quantify. All labour and parts costs must be accurately recorded.

Commissioning failures. These failures can be particularly costly where delays result in deadlines not being satisfied. In addition, revenue can be lost. Specialist labour is often needed for installation, during which there will also be rework, spares and testing costs.

Fault finding in test. Production personnel will often 'bunch' products which have failed test and, if easily corrected faults are found, will correct them. However, where less easily corrected defects are encountered, they may be attended to at the time. This latter type needs to be included under this heading and the cost of investigating these faults must be carefully recorded.

8.1.2 External Assurance Quality Costs

An example of external assurance is an independent testing organisation which tests the product against its specifications. The use of such facilities is becoming increasingly common with high technology products, using computer-based emulators and simulators in the test environment to accurately mimic the real world. Independent testing is vital to qualify a product which, if it fails, could have product liability implications for its producer. These costs will be included in qualification (prevention) in Table 8.1.

Another example of external assurance is an independent assessment organisation. It would perform a detailed and in-depth study of all areas of a company's activities which could affect the quality of its products or services. This is part of audit (prevention) in Table 8.1. Organisations in the United Kingdom who offer this service include British Standards Institution, Lloyd's Register Quality Assurance and the Ministry of Defence (see Chapter 9).

8.2 INTRODUCING A QUALITY COST SYSTEM

Only by knowing where these quality costs are incurred and by recording them, as far as is possible, can management monitor and control them. Quality costs need to be separately collected and recorded, otherwise they become absorbed and concealed in numerous overheads. Many of the items listed already in this chapter can be obtained from a cost accounting system (e.g. rework, test, calibration). Many items, however, will not normally be recorded (e.g. scrap, retest, qualification) and special arrangements will have to be made to record or assess them.

Regular reports to management are vital if there is to be management visibility into quality-related costs. This type of accounting is referred to as memo-accounting since it is over and above traditional accounting practice and is to provide a management 'picture' rather than precise financial data.

It is thus permissible to sample or estimate in order to complete the quality cost report. An estimate is better than no figure at all. This could be obtained by costing a representative percentage of, say, scrap notes and scaling-up to obtain a total.

The steps to be gone through in successfully introducing a quality cost system in a traditional engineering and manufacturing environment are well described in a British Standards Institution standard (BS 6143). It recommends that the first step is a pilot study to determine the scope of work and to identify quality cost types. There are then comments on these, and a checklist.

When the quality cost types have been identified, the collection of data against them can begin. The following steps are suggested; (a) and (b) involve both prevention and appraisal whereas (c) to (e) are failure costs:

(a) Step one is to calculate those costs which are directly attributable to the quality function.

(b) Step two is to identify costs that are not directly the responsibility of the quality function (e.g. stores, purchasing) but which should be included as part of the total quality costs of the company.

(c) Step three is to identify internal failure costs for which budgets were allocated, e.g. planned over-production runs where failures were anticipated.

(d) Step four is to identify internal failure costs for unplanned failures, e.g. reworks, scrap.

(e) Step five is to identify the cost of failures after the change of ownership, e.g. warranty claims.

Quality cost data is varied and will originate from many sources, from payroll/time sheet analysis to material review boards. For the data to be collected consistently, tabulated data sheets are needed, addressing each of the quality cost types identified.

As mentioned above, BS 6143 describes the introduction of a quality cost system into a traditional engineering and manufacturing environment. It has limitations when applied to a high technology/computer-based environment.

8.3 ESCALATING COSTS

Quality costs escalate where there is a lack of management commitment to quality and this attitude flows down through the organisation to the shop-floor. As one moves through the life-cycle, the cost of correcting defects and failures increases dramatically. In an electronics manufacturing company, with machining, assembly, wiring and functional test activities, the following relative cost units will be incurred for defects found at various life-cycle phases:

   (i) A component at incoming inspection and before it is used in engineering: 1 cost unit.
  (ii) The same component used in detailed engineering design pre-production prototypes: 10 cost units.
 (iii) This component in the integrated product, undergoing system testing: 100 cost units.
 (iv) Component failure when the product is in operational use in the field: 1000 cost units.

Thus, when the defect is found during field operation, the failure costs are punitive.

Companies using the most up-to-date technology (e.g. computer-aided design, computer-integrated manufacturing) still find the cost of failure to be crippling. It is not uncommon in the high technology industries that for every 1000 cost units spent on development 700 cost units are then spent on maintaining the item in the field. Even the foregoing quality costs can be overshadowed if there is an omission by upper management to think through the consequences of new policies.

The lessons of the cost of poor quality do not seem to have been learnt. It was estimated, in 1978, by the Department of Prices and Consumer Protection that the turnover for UK industry was £105 thousand million. Since quality costs were estimated to be in the range 4% to 15%, the average was assumed to be 8%, thus involving some £8·4 thousand million.

As failure costs are approximately 50% of the total quality cost, it was costing industry £4·2 thousand million in failures and defects. A 12·5%

reduction in failures would have released £500 million into the economy over this period.

Several research studies have since been funded by the UK Government on the economics of quality control practices and the implementation of statistical process control. The findings were not encouraging. The main conclusions were that there was a lack of understanding of how the practices and processes could reduce quality costs. There was little management commitment and inadequate attention to training.

Software now has an impact on the operational and administrative functions of many companies. Software quality-related costs are huge. A Price Waterhouse study, in 1988, for the Department of Trade and Industry reported that:

> Poor software quality results in substantial costs to both suppliers and users. These 'failure costs' include the costs of correcting errors both before and after delivery of the software as well as costs of overruns and unnecessarily high maintenance costs.
>
> At a conservative estimate, UK users and suppliers currently suffer failure costs of over £500 million a year. This figure includes only domestically produced and marketed software. If one were to include imported software, failure costs would be much higher. There are, in addition, substantial indirect costs that one is unable to quantify. These costs represent the potential benefits to be achieved from improving software quality.

If one could quantify the other areas mentioned plus any costs incurred due to hardware/electronics, the high technology industries alone probably have quality-related costs of well over a billion pounds each year.

Chapter 9

Obtaining Quality System Standards Approvals

9.1 THE NEED FOR QUALITY MANAGEMENT SYSTEM APPROVAL

Industry is well aware of the need to ensure quality—it is vital to efficient cost-effective production. Perhaps more importantly, many major purchasers are now demanding proof of a company's quality record and, as a result, insist that their subcontractors and suppliers operate to a recognised quality management system such as ISO 9000/BS 5750. Examples of such purchasers are British Nuclear Fuels (BNFL), British Gas, British Coal, British Steel, British Telecom, the Ministry of Defence, major motor manufacturers and many well-known companies in the consumer (high street) sector.

It is no longer sufficient to express confidence in one's own quality. The time has come when an independent assessment can be claimed. UK Government-accredited third party organisations, such as BSI, LRQA (Lloyd's Register of Quality Assurance) and Bureau Veritas, can provide that independent assessment. Third party assessments involve an assessor who is contractually independent of both supplier and purchaser.

In the information technology industry many computer system development departments and independent service bureaux are considering updating their quality systems and applying for independent assessment by recognised third parties.

The principal benefits, to industry, of such accreditation are a reduction in multiple assessment which leads to improved cost effectiveness and the growth of a standard form of quality management system, which is more cost effective for industry.

Second party assessments, whereby the purchaser conducts his own assessment, are carried out by the Ministry of Defence.

9.2 PREPARING FOR ASSESSMENT

The work involved in preparing for assessment should not be under-estimated. To have any chance of success the preparation should be treated as a project and all management must be totally committed. It is dangerous to leave all the preparations to a single quality manager. A project team with the required cross-functional authority should be formed.

The first step is to review the existing management system and to assemble all the relevant documentation. Then a comparison should be made with the requirements of BS 5750 and any of the relevant supporting guidelines (see Chapter 2). This will help to identify areas which require changing or where new procedures need to be introduced. This can lead to much heart searching, within the company, to establish a practical and effective quality system to meet the assessment requirements.

A decision should be made on the key areas which have a direct impact on the quality of the operation and the quality system should then be built on these. Once agreement has been reached, the next step is to develop solutions and to implement the appropriate ones.

To support the elements of the quality system, a documentation package has to be produced. This will take the form of a quality manual, supported by procedures as described by this book. It is important to stress that the documentation must reflect the actual system in action.

9.3 THIRD PARTY ASSESSMENT

Both BSI and LRQA carry out assessments on companies, wishing to be accredited to BS 5750 Part 1, 2 or 3 1987. Initially BSI or LRQA would issue, on request, a customer information questionnaire which asks questions relating to:

—company policy
—number of employees
—details of the product or service offered
—type of market and customers
—details of quality procedures/system
—details of any quality approvals
—general system of documentation in use

On receipt of the questionnaire, any questions from the company, concerning the assessment process, are answered. Since the scope and criteria for assessment vary greatly from company to company it is not practicable to discuss assessment costs here. This information will always be readily provided by the assessor according to the circumstances and the rates prevailing at the time.

9.3.1 BSI/LRQA Assessment

On completion of formalities, a contract is signed and the first stage involves a review of the quality documentation. In the first instance BSI will wish to see the quality manual, this being the top of the quality documentation hierarchy. LRQA will wish to see both the manual and the quality documentation at this stage.

One purpose of the review is to ensure that all the necessary documentation to meet the requirements of BS 5750/ISO 9000 is in existence, that it is controlled and distributed. If documentation is not distributed it effectively does not exist. This is a minimum requirement and there is no point in incurring the cost of even a preliminary assessment unless this is already in place. The effectiveness of the documents cannot be assessed at this stage. That is addressed during the review of the implementation of the various procedures.

The next step is to establish the scope and criteria of the assessment. This is done against the appropriate part of BS 5750/ISO 9000 for which accreditation is sought. In addition, guidance documents such as BS 5750 Part 4, which enlarge on the sections of Parts 1, 2 and 3, will be applied. In the case of BSI, QASs (quality assessment schedules) are also used and, for LRQA, QSSs (quality system supplements) are used. These are industry-specific interpretations according to the products and services. Where these are not applicable, BSI operates the QUASAR (quality assessment and registration) scheme, whereby a firm is assessed without any specific industry guidance and to a scope and operation agreed with BSI.

Any product-specific or statutory requirements and standards are also considered at this stage. A tour of the premises is then conducted and a programme (schedule) for the assessment is established.

The assessment team then visits the company to examine the operation of the quality management system described in the manual and procedures. These are compared with the BS or LRQA criteria mentioned above. Objective evidence is sought that the system is in effective operation.

The assessor will be primarily interested in the records that result from the quality procedures since it is only these that provide the evidence of the quality system implementation and effectiveness. In particular, records of defects and of customer complaints, including changes in work instructions that have resulted from corrective actions, need to be available since they provide solid records of an evolving and practical quality system.

If any aspects of the operation do not meet the requirements of the scheme, the deficiencies are noted on discrepancy reports (BSI) or non-compliance notes (LRQA). BSI, at the end, issue a summary report with recommendations for remedial action. LRQA issue an 'assessment summary and recommendations'.

Should there be the need for remedial action, then a suitable completion date will be agreed and, providing that the corrective actions are carried out within the agreed time, the reassessment will only address those parts of the quality system needing remedial action.

On completion of a satisfactory assessment the company will be issued with an approval certificate which describes the extent of the approval. There is no time limit to the BSI approval but they reserve the right to carry out four unannounced surveillance visits each year. If deficiencies are found, and not corrected, the certificate can be revoked. The LRQA certificate is valid for three years subject to satisfactory maintenance of the quality management system.

When a registered company changes its scope of operations then a part or whole reassessment of the quality management system will be needed.

Successful registration permits a company to broadcast the fact in its advertising, marketing literature, letterheads and so on. By carrying the assessment logo a company gives itself a marketing and competitiveness edge. There is also an LRQA and a BSI quality mark which can be carried by registered companies.

9.4 SECOND PARTY—MOD ASSESSMENT

If the Ministry of Defence is the customer then it is likely to carry out its own assessments against MOD standards; AQAP 1 for quality control requirements for industry and AQAP 13 for software quality control system requirements (see Chapter 2). To obtain MOD (AQAP) approval, companies are required to show evidence of having dealt directly with MOD in the past. Alternatively, they must find an MOD Procurement Executive Contracts branch who will act as their sponsor. The sponsor states, in effect, that he will be prepared to place an order if the company obtains the appropriate level of approval.

MOD questionnaires are then obtained by prospective suppliers. These seek information about the company's structure and quality system. A copy of the last two years' accounts must be provided.

An MOD assessor will make a preliminary visit to the company and will suggest any areas needing attention. MOD approval, when obtained, is for three years with surveillance visits during that period.

Unlike BSI and LRQA, the MOD makes no charge for assessment since it is evaluating its own suppliers, as required by its own quality assurance system.

9.5 USEFUL ADDRESSES

British Standards Institute application forms can be obtained from:

BSI,
Business Development Unit,
Certification and Assessment,
PO Box 375,
Milton Keynes MK14 6LL
Tel: 0908 220908

Lloyd's Register application forms can be obtained from:

Lloyd's Register,
Customer Services,
Norfolk House,
Wellesley Road,
Croydon,
Surrey CR9 2DT
Tel: 081 688 6882

Bureau Veritas application forms can be obtained from:

Bureau Veritas,
Challenge House,
Sherwood Drive,
Bletchley,
Milton Keynes MK3 6DP
Tel: 0908 366724

MOD/AQAP application forms are available from:

DGDQA (Director General Defence Quality Assurance),
Woolwich Arsenal,
London SE18 6TD

PART 4

Sample Quality Manuals

COMPANY A OR COMPANY B
QUALITY MANUAL

Compiled by...

Checked by...

Approved by..

Date...

Note: The term Company A is used throughout this example. It refers equally to Company B.

This manual is registered in your name to enable the quality department to update the contents when required.

Please notify the quality department if the name or the title of the holder are incorrect, or if there is any other change.

This manual must be returned to the engineering and quality manager when the holder leaves the department.

Under no circumstances may copies be made.

Please sign the acknowledgement below and return the attached copy to the engineering and quality manager.

NAME ....

TITLE ....

QUALITY MANUAL NUMBER ....

ISSUE ....

DATE ....

CHANGE HISTORY PAGE

Manual status	Issue control	Date issued	Number of pages	Changed pages	Change/ defect no.
Controlled	XX	XX	XX	XX	XX

QUALITY MANUAL

FOREWORD

The prosperity of Company A can only be assured by the continued satisfaction of its customers. Product quality and reliability are hence essential elements of this objective.

The company has some years of experience in design and manufacture, assembly and operation of its products. The level of success achieved to date can only be maintained by a continuing programme of quality and reliability improvement.

Whilst quality procedures can be specified, their effectiveness depends upon the attitudes of personnel at all levels within the organisational structure. The aim, therefore, is to foster quality awareness by all the means at our disposal.

DECLARATION

This quality manual is devoted to describing the activities within Company A, the aims of which are to provide products and services to satisfy the standards and requirements of our customers. The objective of the company quality assurance system is to deliver the agreed product through a policy of set procedures operated throughout the entire organisation.

The provisions of this quality manual have been reviewed by me and I certify that it will be used as a working document, enforced by the engineering and quality assurance manager, who shall have the necessary authority for ensuring that the requirements are implemented and maintained.

Signed ....

MANAGING DIRECTOR

QUALITY MANUAL

CONTENTS

0 INTRODUCTION

The Company A quality assurance procedures and work instructions have been compiled to ensure that products are designed and manufactured to the requirements of BS 5750 (Part 1) 1987 (ISO 9001: 1987).

The responsibility for ensuring compliance with contract requirements and standards has been delegated to the engineering and quality manager by the managing director.

Section 4 of this manual is numbered using the identical paragraph numbers to the requirements in BS 5750 (Part 1) 1987.

0.1 DISTRIBUTION AND UPDATING OF THE MANUAL

The engineering and quality manager, or his appointed deputy, is responsible for distributing and updating the manual. The engineering and quality manager alone is responsible for the administration and the raising and incorporation of all amendments. Each holder of the quality manual is responsible for maintaining and keeping his/her copy up to date when revisions are issued.

All copies of the manual shall be returned to the engineering and quality manager when a registered holder resigns, or for any other reason has no further need of it.

The engineering and quality manager may distribute 'uncontrolled copies' of the quality manual to external companies. These copies will not be updated. He will periodically review and revise the quality system.

1 PURPOSE AND SCOPE

This quality manual describes and designates the management responsibilities and procedures in design, planning, procurement, manufacture, inspection, test and quality that will be applied by Company A to meet the requirements of BS 5750 (Part 1) 1987 (ISO 9001: 1987).

The contents of this manual are mandatory and must not be altered or omitted without the written authority of the engineering and quality manager.

2 APPLICABLE DOCUMENTS AND REFERENCES

(1) BS 5750 (Part 1) 1987 (ISO 9001: 1987) Quality systems—specifications for design/development, production, installation and servicing.

(2) BS 4778: 1987 (ISO 8402: 1986) Quality vocabulary—international terms.

(3) Company A quality procedures and work instructions referred to in the sections of this document and listed in Sections 5 and 6 respectively.

(4) BS 6000 Guide to the use of BS 6001.
BS 6001 Sampling procedures and tables for inspection by attribute.

(5) BS 6200 Series: soldering and board assembly.

3 DEFINITIONS

For the purposes of this manual, the definitions given in BS 4778 (Part 1) 1987 (ISO 8402: 1986) shall apply.

4 REQUIREMENTS

4.1 MANAGEMENT RESPONSIBILITY

4.1.1 Company Quality Policy

The objective of the management policy is to ensure that products and services supplied by the company are fit for their intended purpose, ensuring performance, safety and reliability in operation. This policy is implemented and understood at all levels within the organisation through well-defined company procedures and work instructions which operate and maintain a quality system in compliance with BS 5750 (Part 1) 1987 (ISO 9001: 1987).

4.1.2 Organisation

4.1.2.1 Responsibility and Authority

Company A is a compact progressive company staffed by a blend of experienced personnel skilled in the design and manufacture of components and equipments.

The managing director has complete authority over all matters pertaining to quality. Through the engineering and quality manager, he has delegated the day-to-day responsibility for ensuring that procedures, work instructions and documentation used by any personnel of Company A are

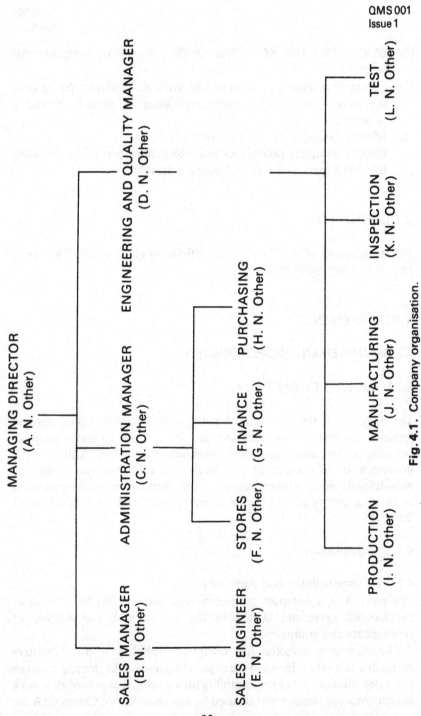

Fig. 4.1. Company organisation.

implemented and coordinated so that the products and services supplied are in full compliance with contract and quality requirements.

The organisation of the company is shown in Fig. 1. Job descriptions are held by the engineering and quality manager.

4.1.2.2 Verification Resources and Personnel

The requirements for in-house verification of design, procurement, manufacture, testing, installation and servicing of all Company A products are identified in the procedures and work instructions.

For a list of project procedures and work instructions see Sections 5 and 6 respectively.

Inspections, internal audits and design reviews are carried out by personnel independent of those responsible for the work being performed.

4.1.2.3 Management Representative

The engineering and quality manager is the appointed representative of the company for all matters pertaining to product and service quality. He has invested in him the authority and responsibility for ensuring that the company quality policy is maintained.

4.1.3 Management Review

The managing director shall review the quality system by assessing the results of internal quality audits and by arranging third party audits. The quality system shall be reviewed at least once a year.

There is no separate procedure for this item but the managing director will maintain a file of planned and past reviews together with details of results and corrective actions.

4.2 QUALITY SYSTEM

4.2.1

The elements of the company quality system are:

(1) Applicable national standards.
(2) Formalised company procedures.
(3) Company work instructions.

The effectiveness of the quality system is the responsibility of the engineering and quality manager acting on behalf of the managing director for the:

(a) Coordination and monitoring of the quality system.
(b) Resolution of any non-conformance in the system.
(c) Implementation of effective actions to be taken by appropriate personnel to ensure compliance with specific requirements.
(d) Updating of quality procedures, work instructions, specifications, processes and testing techniques, as necessary.
(e) Identification and preparation of quality records.
(f) Internal quality audits.

4.3 CONTRACT REVIEW

A contract is defined as an agreement, verbal or written, to perform a service or provide a product, at some future time, for a financial consideration.

It is the responsibility of the sales manager to initiate reviews of any activities which bear on the ability of the company to meet the requirements of each order. He will involve the engineering and quality manager in the review of each contract involving non-standard parts.

Sales and contract control is covered by

QPS 001—Contract Review

4.4 DESIGN CONTROL

4.4.1 General

Compliance is achieved by verifying that the design criteria comply with specific requirements so that data and methods are valid for the range of application and that completed designs satisfy design criteria. The implementation of company procedures is the mechanism by which control of design is achieved.

4.4.2 Design and Development Planning

Each product design and development is the subject of a design sheet agreed prior to the contract commencing.

This will be done for both Company A products and any subcontracted products.

4.4.2.1 Activity Assignment

Design and verification activities are the responsibility of the engineering and quality department to ensure that adequate resources are provided so as to permit effective execution of the development.

4.4.2.2 Organisation and Technical Interface

Technical interfaces will be at manager level.

4.4.3 Design Input

Design requirements are clearly identified by discussions with the customer. These are documented and agreed by means of the design sheet. Incomplete or conflicting requirements of the contract are to be resolved, where possible, before proceeding with the task.

4.4.4 Design Output

Procedures are implemented to review the design for correctness to specification and performance. Also that specific requirements for quality have been met when measured against any appropriate regulatory codes, procedures and standards.

4.4.5 Design Verification

Original design data is independently verified for compliance with specific criteria and any project work is documented and maintained in a project file.

The extent of design verification is dependent upon the features, complexity and requirements of the item under consideration. The design method, wherever possible, uses authenticated techniques to minimise the probability of errors and uncertainties, by utilising established practices and standards, and maintaining extra care in any extrapolation beyond previous experience.

The verification of the design of new equipment is established by prototype testing and detailed analysis of the performance of first-off production units. Service life performance may be sought from the use of accelerated life test programmes if deemed necessary by the engineering and quality manager.

Design reviews are carried out at predetermined stages in order to conduct a formal examination of all aspects of design and development;

also to verify that technical and quality control requirements are being met and that the necessary corrective actions are being taken where necessary.

Design is controlled by

> QPS 016—Calibration of Measuring Equipment
> QPA 002—Configuration and Change Control (or QPB 002)
> QPS 003—Design and Review
> QPB 004—Software Development and Standards (Company B only)
> QPS 001—Contract Review
> WIS XXX—Engineering Drawing

4.5 DOCUMENT CONTROL

4.5.1 Documentation Approval and Issue

The routines established cover all documents related to the performance and quality of work in progress such as procedures, work instructions, production drawings, certificates and general office administration. The control of such documents is a task which is the responsibility of all, in order to authenticate the work undertaken.

The preparation, issue and change of documents are controlled to ensure that correct documents are being employed. Such documents, including changes thereto, are reviewed for adequacy and approval for release by authorised personnel.

The approval of documents that form part of the final documentation will be a continuous process throughout any given contract.

4.5.2 Document Changes

Proposed changes within the content of particular contracts, initiated by either the company or its clients, are agreed by the sales and engineering and quality managers. A formal written communication, between both parties, on the nature of the change proposal and the action to be taken, is then sent.

Pertinent number issues of appropriate documents are then brought under internal change control. The internal control of drawing issues is through a document register. Details of component items and their drawing issues are detailed therein.

Document control is controlled by

> QPA 002—Configuration and Change Control (also QPB 002)

QPA 008—Documentation Format and Standard (also QPB 008)
WIS XXX—Archiving

4.6 PURCHASING

4.6.1 General

The purchase of material, components and services is the responsibility of the administration manager, who shall endorse purchase requisitions raised by stores. Where necessary a quotation will first be requested.

4.6.2 Assessment of Subcontractors

All sources of supply must be evaluated and approved by the engineering and quality manager prior to the placement of the order. The type and extent of the evaluation shall depend on the nature of the goods or services to be provided and the degree of previous experience with the supplier.
Where possible suppliers shall be BS 5750 registered.
To enable the engineering and quality manager to carry out an on-site supplier evaluation, a checklist is available. Suitable suppliers will be placed on an approved suppliers' list. The engineering and quality manager shall be responsible for ensuring that the records of suppliers' performances are kept and that regular assessments are made. A history of all rejected goods will be maintained in respect of individual suppliers and this will be used to monitor their performance.
Suppliers regularly appearing on the rejected list are first advised that they must improve their quality. If no improvement is noted their names will be removed from the Approved Suppliers' List.

4.6.3 Purchasing Data

All purchase orders will contain adequate data to enable the supplier to provide the correct quantity and quality of material or services.
All purchase orders will be reviewed and signed by the managing director.

4.6.4 Verification of Purchased Product

Facilities are provided at Company A for customers to witness products at any stage of manufacture.

Purchased goods are controlled by

QPS 009—Purchasing and Suppliers
QPA 011—Goods Inwards and Inspection (or QPB 011)
QPS 007—Subcontracted Equipment and Software (Company B only)
QPS 022—Statistical Techniques

4.7 PURCHASER SUPPLIED PRODUCT

In some instances the customer may supply free issue material which will be treated as a purchased item and will follow the appropriate procedures, including inspection, storage and maintenance.

If any such material is lost, damaged or otherwise unsuitable it shall be recorded and reported to the purchaser.

This is controlled by

QPS 009—Purchasing and Suppliers
QPA 011—Goods Inwards and Inspection (or QPB 011)
QPS 022—Statistical Techniques

4.8 PRODUCT IDENTIFICATION AND TRACEABILITY

All material that has passed the goods inwards inspection is transferred to the appropriate store, segregated and individually identified, where possible.

All packages containing products are marked with appropriate product identification labels, when in the manufacturing areas. Cartons or boxes are marked with similar labels for identification. Large items of equipment also have labels attached for identification.

This is controlled by

QPA 010—Control of Production (or QPB 010)
QPS 012—Identification and Traceability
QPS 013—Stores

4.9 PROCESS CONTROL

4.9.1 General

Manufacturing, assembly and testing operations are performed under controlled conditions. These conditions include drawings and work

98

instructions, production and test procedures, used during and on completion of assembly. Testing and inspection procedures and quality criteria, which define the method and sequences of all manufacturing operations, are provided.

4.9.2 Special Processes

Production data is collected for reporting the long-term performance of products installed as operational systems. Appropriate records are maintained for analysis by design and manufacturing.

4.10 INSPECTION AND TESTING

4.10.1 Receiving Inspection and Testing

All goods that arrive are received at goods inwards, where they will be inspected by appropriate personnel.

Goods will only be accepted against the purchase requisition, relevant drawings, specifications and other documentation.

All received goods will be checked for quantity and conformance to the purchase requisition as described in the appropriate procedures.

All goods inwards will be suitably colour-marked to indicate conformance or non-conformance and kept separately in the goods inwards or stores areas to await use or further action.

4.10.2 In-process Inspection and Testing

During the course of product assembly, inspections and tests are carried out in accordance with applicable documented procedures, drawings production and clients' specifications, etc. The results of inspections and tests are recorded on production documents attached to each batch of manufactured items to cover various stages in their assembly. Non-conforming items are identified, noted on the documents, and reported to the engineering and quality manager for segregation and appropriate action. Within each batch all items are individually identified on the documents.

4.10.3 Final Inspecting and Testing

All assemblies are tested functionally prior to despatch. Such tests can be witnessed by the client. Test processes and acceptance criteria are marked on the production documents. These are agreed with the client before any testing is carried out. On satisfactory completion of testing, where necessary, acceptance certificates are signed by the client and by Company A.

4.10.4 Inspection and Test Records

Material certificates, manufacturing data, drawings, test results, etc., are all collated as a contract progresses and are maintained in the production folder. At the end of the contract the file will be archived.

Inspection and testing are controlled by

QPS 016—Calibration of Measuring Equipment
QPA 011—Goods Inwards and Inspection (or QPB 011)
QPA 017—Non-conformance, Corrective Action and Records (also QPB 017)
QPA 014—Line Inspection and Test (also QPB 014 and QPB 015)
QPS 022—Statistical Techniques

4.11 INSPECTION, MEASURING AND TEST EQUIPMENT

All inspection and test equipment used to verify compliance is subjected to regular calibration checks where applicable.

Each instrument is allocated a unique reference number and is checked against master references which are traceable to national standards. Where equipment cannot be calibrated in-house, an approved testing house which is a member of the British Calibration Service is utilised.

A label is fixed to each instrument stating the date on which it was last calibrated. Records of calibration results are maintained in the calibration register held by the engineering and quality manager.

This is controlled by

QPS 016—Calibration and Measuring Equipment

4.12 INSPECTION AND TEST STATUS

The status of items at the goods receiving and inspection stage is referenced in Section 4.10.

The inspection status of items during assembly is recorded in the production folder, one of which is attached to each batch of manufactured items as work progresses to final assembly. Each item within a batch may be uniquely identifiable as appropriate.

Completed items are not released until final inspection and testing has been carried out, sometimes in the presence of the client. The production folder will record details of inspection and tests. In the case of customer-witnessed inspection the signature of the customer on the acceptance certificate (or release note) shall constitute acceptance that the product conforms with its specification.

This is covered by

> QPA 010—Control of Production (or QPB 010)
> QPA 014—In-line Inspection (or QPB 014)

4.13 CONTROL OF NON-CONFORMING PRODUCT

Those actions pertaining to the control of non-conforming products are located at goods inwards inspection, where appropriate closed-off or locked shelves exist.

Additionally, items which fail within this non-conforming category during assembly and testing of the manufactured product are similarly segregated and actioned by the engineering and quality manager.

In the latter case, the production files shall be completed, indicating which units of a batch failed and at which point of assembly.

This is covered by

> QPA 017—Non-conformance, Corrective Action and Records (also QPB 017)

4.14 CORRECTIVE ACTION

At various stages of a contract (e.g. design, purchasing and manufacture) it is sometimes necessary to initiate corrective actions to ensure compliance with standards, procedures and specifications. The corrective actions are documented in the procedures and work instructions, and cover the investigation and analysis of faults, fault reporting and repairs to items returned from site.

This is covered by

> QPA 017—Non-conformance, Corrective Action and Records (also QPB 017)

4.15 HANDLING, STORAGE, PACKAGING AND DELIVERY

4.15.1 General

Items and completed assemblies are moved around the workshop manually.

4.15.2 Handling

Staff are instructed in the safe handling of items so as to prevent their deterioration.

4.15.3 Storage

Secure storage areas are provided for all work in progress, where the areas are protected to prevent deterioration of stock.

4.15.4 Packaging

Items in stores are kept on shelves marked with the appropriate labels. These identification labels should be transferred to the cartons or boxes in which the items are packed.

4.15.5 Delivery

Preservation of completed assemblies for despatch and delivery is sometimes stipulated by clients in the contract specification. Packaging and crating of completed items is in accordance with contract specifications. In some cases specialist carriers are used to convey items to clients.

This is covered by

QPS 013—Stores
QPS 022—Statistical Techniques

4.16 QUALITY RECORDS

Records are maintained which verify the effective operation of the quality system. These records include drawings, specifications, procedures, production folders, certificates, reports, corrective actions, inspection results, calibration results and test reports. The records for each contract are collated in a project file, a copy of which is maintained in the Company

A archive for a minimum of 3 years, or longer if the contract stipulates. During the retention period clients shall have access to the project files in the archives.

Most of the quality system documents are applicable. This is covered by

QPS 018—Quality Records

4.17 INTERNAL QUALITY AUDITS

Internal quality audits are carried out to verify that quality activities comply with the planned arrangements and also to determine the effectiveness of the quality system.

Audits are carried out on a regular and planned basis. The results and corrective actions are notified to the managing director. There shall be a re-audit within a specified period to determine the status of the corrective action.

There is a separate procedure for this item and the engineering and quality manager will maintain a file of planned and past audits together with details of results and corrective action.

This is covered by

QPA 019—Internal Quality Audit (also QPB 019, which includes subcontractor audits)

4.18 TRAINING

Manpower resources and facilities in the company are periodically and systematically reviewed against past, present, planned and forecast levels of business activity by product type, volume and mix, to determine and regulate the forward programmes of recruitment and training from which future manpower resources will be met.

The managing director is responsible for specifying the minimum entry requirements, appraisal of applicants, and selection and induction of the new entrant. He is also responsible for identifying and satisfying the training needs of all existing personnel in their areas of operation.

Training programmes which comply with the requirements of appropriate external bodies, such as the Engineering Industry Training Board, can be used to satisfy training requirements.

This is covered by

QPS 020—Training

4.19 SERVICING

Contracts sometimes include a requirement for servicing and maintenance whereby the quality of the product is maintained.

The duties called for within a service contract would be performed by suitably qualified personnel having the required skills and experience.

This is covered by

QPS 021—Servicing

4.20 STATISTICAL TECHNIQUES

Procedures will be established to verify the acceptability of a product and any processes.

These procedures will establish statistical techniques, quality measures and their implementation. The data used in the quality measures will be collated and collected against a product specification in terms of its conformance to its specification and may include:

(1) reliability
(2) accuracy
(3) precision

This data can only be meaningfully interpreted when sufficient records are available. Trend analysis will then be carried out.

The relevant document is

QPS 022—Statistical Techniques

5 LIST OF QUALITY PROCEDURES

Procedure number	Title
QPS 001	Contract Review
QPA 002	Configuration and Change Control (and QPB)
QPS 003	Design and Review
QPB 004	Software Development and Standards (Company B only)
QPS 007	Subcontracted Equipment and Software (Company B only)
QPA 008	Documentation Format and Standard (and QPB)

QPS 009	Purchasing and Suppliers
QPA 010	Control of Production (and QPB)
QPA 011	Goods Inwards and Inspection (and QPB)
QPS 012	Identification and Traceability
QPS 013	Stores
QPA 014	Line Inspection and Test (and QPB 015)
QPB 014	In-line Inspection
QPB 015	Test
QPS 016	Calibration of Measuring Equipment
QPA 017	Non-conformance, Corrective Action and Records (and QPB)
QPS 018	Quality Records
QPA 019	Internal Quality Audit (and QPB, which includes subcontractor audits)
QPS 020	Training
QPS 021	Servicing
QPS 022	Statistical Techniques

6 LIST OF WORK INSTRUCTIONS

WIS XXX	Engineering Drawing
WIS XXX	Archiving
WIS XXX	Packaging and Despatch
WIS XXX	General Housekeeping
WIS 501	Widget Test Procedure

and so on as applicable to Company A or Company B.

COMPANY C
QUALITY MANUAL

Compiled by...

Checked by ....

Approved by...

Date...

This manual is registered in your name to enable the technical department to update the contents when required.

Please notify the technical department if the name or the title of the holder are incorrect, or if there is any other change.

This manual must be returned to the technical director when the holder leaves the department.

Under no circumstances may copies be made.

Please sign the acknowledgement below and return the attached copy to the technical director.

NAME ....

TITLE ....

QUALITY MANUAL NUMBER ....

ISSUE..

DATE ....

CHANGE HISTORY PAGE

Manual status	Issue control	Date issued	Number of pages	Changed pages	Change/ defect no.
Controlled	XX	XX	XX	XX	XX

QUALITY MANUAL

FOREWORD

The prosperity of Company C can only be assured by the continued satisfaction of its customers. Product quality and reliability are hence essential elements of this objective.

The company has many years of experience in the software industry and the level of success achieved to date can only be maintained by a continuing programme of quality and reliability improvement.

Whilst quality procedures can be specified, their effectiveness depends upon the attitudes of personnel at all levels within the organisational structure. The aim, therefore, is to foster quality awareness by all the means at our disposal.

DECLARATION

This quality manual is devoted to describing the activities within Company C, the aims of which are to provide products and services to satisfy the standards and requirements of our clients. The objective of the company quality assurance system is to deliver the agreed product or service through a policy of set procedures operated throughout the entire organisation.

The provisions of this quality manual have been reviewed by the board of directors and I certify that it will be used as a working document, enforced by the technical director, who shall have the necessary authority for ensuring that the requirements are implemented and maintained.

Signed ....

MANAGING DIRECTOR

QUALITY MANUAL

CONTENTS

0 INTRODUCTION

The Company C quality assurance procedures and work instructions have been compiled to ensure that products are designed and provided to the requirements of BS 5750 (Part 1) 1987 (ISO 9001: 1987).

The responsibility for ensuring compliance with contract requirements and standards has been delegated to the technical director by the managing director.

Section 4 of this manual is numbered using the identical paragraph numbers to the requirements in BS 5750 (Part 1) 1987.

0.1 DISTRIBUTION AND UPDATING OF THE MANUAL

The technical director, or his appointed deputy, is responsible for distributing and updating the manual. The technical director alone is responsible for the administration and the raising and incorporation of all amendments. Each holder of the quality manual is responsible for maintaining and keeping his/her copy up to date when revisions are issued.

All copies of the manual shall be returned to the technical director when a registered holder resigns, or for any other reason has no further need of it.

The technical director may distribute 'uncontrolled copies' of the quality manual to external companies. These copies will not be updated. He will periodically review and revise the quality system.

1 SCOPE

This quality manual describes and designates the management responsibilities and procedures in the technical, support, sales and administration departments that will be applied by Company C to meet the requirements of BS 5750 (Part 1) 1987 (ISO 9001: 1987).

The contents of this manual are mandatory and must not be altered or omitted without the written authority of the technical director.

2 REFERENCES

(1) BS 5750 (Part 1) 1987 (ISO 9001: 1987) Quality systems—specifications for design/development, production, installation and servicing.
(2) BS 4778: 1987 (ISO 8402: 1986) Quality vocabulary—international terms.

(3) Company C quality procedures referred to in the sections of this
document and listed in Sections 5 and 6 respectively.

3 DEFINITIONS

For the purposes of this manual, the definitions given in BS 4778 (Part 1)
1987 (ISO 8402: 1986) shall apply.

4 REQUIREMENTS

4.1 MANAGEMENT RESPONSIBILITY

4.1.1 Company Quality Policy

The objective of management policy is to ensure that products and services
supplied by the Company C are fit for their intended purpose, ensuring
safety and reliability in operation. This policy is implemented and
understood at all levels within the organisation through well-defined
company procedures which operate and maintain a quality system in
compliance with BS 5750 (Part 1) 1987 (ISO 9001: 1987).

4.1.2 Organisation

4.1.2.1 Responsibility and Authority
Company C is based at the address shown on the front sheet. It is a
completely independent company under the direction of Mr A. N. Other.

The technical director has complete authority over matters pertaining to
quality. Through the senior members of the technical support department,
the technical director has responsibility for ensuring that the quality
procedures are implemented by all personnel of Company C. Senior
members of the technical support department are project managers and
senior analysts/programmers.

The organisation of the company is shown in Fig. 1. Job descriptions are
held by the technical director.

4.1.2.2 Verification Resources and Personnel
The requirements for in-house verification of products are identified in the
procedures.

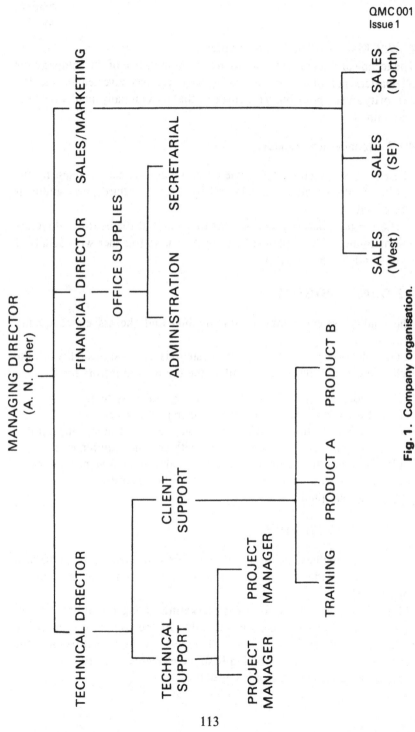

Fig. 1. Company organisation.

4.1.2.3 Management Representative

The technical director is the appointed representative of the company for all matters pertaining to product quality. He has invested in him the authority and responsibility for ensuring that the company quality policy is maintained.

4.1.3 Management Review

The board of directors shall review the quality system by assessing the results of internal quality audits and by arranging third party audits, if appropriate.

There is no separate procedure for this item but the board of directors will maintain a file of planned and past reviews together with details of results and corrective action.

4.2 QUALITY SYSTEM

The quality system comprises this manual and the associated quality procedures.

The effectiveness of the quality system is the responsibility of the technical director acting on behalf of the board of directors for the:

(a) Coordination and monitoring of the quality system.
(b) Resolution of any non-conformance in the system.
(c) Implementation of effective actions to be taken by appropriate personnel to ensure compliance with specific requirements.
(d) Updating of the quality procedures and manual as necessary.
(e) Identification and preparation of quality records.
(f) Internal quality audits.

4.3 CONTRACT REVIEW

A contract is defined as an agreement, verbal or written, to perform a service or provide a product, at some future time, for a financial consideration.

It is the responsibility of the sales/marketing manager to initiate reviews of any activities which bear on the ability of the company to meet the requirements of each contract. He will involve the technical director in the review of each contract involving bespoke elements of work.

Sales and contract control is covered by

QPC 001—Contract Review

4.4 DESIGN CONTROL

4.4.1 General

Compliance is achieved by verifying that the design criteria comply with specific requirements so that data and methods are valid for the range of application and that completed designs satisfy design criteria. The implementation of company procedures is the mechanism by which control of design is achieved.

4.4.2 Design and Development Planning

Each product's design and development follows the requirements of

QPC 004—Software Development
QPC 005—Software Design Standards
QPS 007—Subcontract Equipment and Software

This will be done for both Company C products and any subcontracted products.

4.4.2.1 Assignment of Responsibilities

Design and verification activities are the responsibility of the technical support department to ensure that adequate resources are provided so as to permit effective execution of the development.

4.4.2.2 Organisation and Technical Interface

Technical interfaces will be at manager level and controlled by the use of project progress reports, the use of which is defined in

QPC 004—Software Development

4.4.3 Design Input

Design requirements are clearly identified by discussions with the customer. These are documented and agreed by means of the identification report, proposal, feasibility report or system description. Each report is more detailed and should be agreed with the client before proceeding to the next phase. Incomplete or conflicting requirements of the contract are to be resolved, where possible, before proceeding with the task.

QPC 004—Software Development

4.4.4 Design Output

Procedures are implemented to review the design documents at each stage for correctness to specification and performance.

QPC 006—Software Review

4.4.5 Design Verification

The design verification is carried out by senior members of the technical support department, in accordance with QPC 006, at each stage of a project. All notes and documentation for that project are maintained in a project file.

Design is controlled by

QPC 001—Contract Review
QPC 002—Configuration and Change Control
QPC 004—Software Development
QPC 005—Software Design Standards
QPS 006—Software Review

4.5 DOCUMENT CONTROL

4.5.1 Documentation Approval and Change Control

The document approval and change control are covered by the following quality procedures:

QPC 002—Configuration and Change Control
QPC 004—Software Development
QPC 005—Software Design Standards
QPC 008—Documentation Format and Standard

The above procedures define the format of all documents and their content. The review process is performed on all documents in a controlled state and all must undergo review to become controlled.

4.6 PURCHASING

4.6.1 General

The purchase of material, components and services is the responsibility of the administration manager, who shall process purchase orders raised by

116

any member of staff. All orders not of a general office supplies nature must be approved by a director of the company. Full details of the procedures to be carried out are given in

QPC 009—Purchasing and Suppliers

4.6.2 Assessment of Subcontractor

All sources of supply must be evaluated and approved by a director of the company prior to the placement of the order. The type and extent of the evaluation shall depend on the nature of the goods or services to be provided and the degree of previous experience with the supplier.

Where possible suppliers shall be BS 5750 registered.

4.6.3 Purchasing Data

All purchase orders will contain adequate data to enable the supplier to provide the correct quantity and quality of material or services.

All purchase orders will be reviewed and signed before release, and the level of authority is defined in

QPC 009—Purchasing and Suppliers

Facilities are provided at Company C for customers to witness products at any stage of manufacture.

Purchased goods are controlled by

QPC 009—Purchasing and Suppliers
QPC 011—Goods Inwards, Inspection and Stores
QPS 007—Subcontracted Equipment and Software
QPC 022—Statistical Techniques

4.7 PURCHASER SUPPLIED PRODUCT

In some instances the customer may supply free issue material or software which will be treated as a purchased item and will follow the appropriate procedures, including inspection, storage and maintenance.

If any such material is lost, damaged or otherwise unsuitable it shall be recorded and reported to the purchaser.

This is controlled by

QPC 009—Purchasing and Suppliers
QPC 011—Goods Inwards, Inspection and Stores
QPC 022—Statistical Techniques

117

4.8 PRODUCT IDENTIFICATION AND TRACEABILITY

All goods which have passed the goods inwards inspection are transferred to the appropriate store, segregated and individually identified, where possible. Where these goods are software then the version number must be recorded.

The following paragraph does not apply to office stationery supplies, where it is the responsibility of the office administration manager to ensure that sufficient stock levels are maintained.

All software products will be identified with labels detailing the version number and date, and the label will remain with the software product whilst in the possession of Company C. If the software product is to be released to the client it must first be reviewed and audited in line with

QPC 006—Software Review

No software product in an uncontrolled state can be released to a client for any purpose.

4.9 PROCESS CONTROL

4.9.1 General

Process control involves monitoring and auditing the development cycle of a product. This is covered in

QPC 005—Software Design Standards
QPC 015—Test
QPC 017—Non-conformance, Corrective Action and Records
QPC 019—Internal Quality Audit

4.9.2 Special Processes

All software projects can be regarded as a special process, in that some areas of the system may not cater fully for the requirements of the client as previously defined and agreed with that client. The quality procedure

QPC 017—Non-conformance, Corrective Action and Records

details the process of monitoring and correcting faults.

The review and audit process will identify personnel responsible for software projects, as will the task schedules required by the development process.

4.10 INSPECTION AND TESTING

4.10.1 Receiving Inspection and Testing

All goods which arrive at Company C are received at goods inwards. All products not of an office supply nature will be inspected, verified and tested before being released for use. This applies to computer hardware products and software products. At not time should any software product be released for use without verification. It may be necessary from time to time to release computer hardware prior to inspection. If this is the case then it should be recorded as such in the goods inwards log and then inspected at a later date.

QPC 019—Internal Quality Audit

4.10.2 In-process Inspection and Testing

During the course of system development, inspections and tests are carried out in accordance with applicable documented procedures.

QPC 015—Test

4.10.3 Final Inspecting and Testing

The final inspection and test activity is termed system test and is conducted by a party independent of the development team on the project. This is defined in

QPC 015—Test

4.10.4 Inspection and Test Records

All test plans and test results for individual functions and for whole systems are maintained for all projects.

4.11 INSPECTION, MEASURING AND TEST EQUIPMENT

This does not apply to the scope of Company C activities.

119

4.12 INSPECTION AND TEST STATUS

The inspection and test status and software will be monitored by

QPC 006—Software Review
QPC 002—Configuration and Change Control

The issue of version numbers will be dependent upon the source of the software. If under the control of Company C then version numbers will be issued upon review and after subsequent approval. The approved software will then be protected from any further change on the part of development staff by the technical support department. Software not under the control of Company C (i.e. third party software) will already have version numbers applied and these will be noted in the review and software control procedures.

All hardware products will be controlled by serial numbers, recorded on receipt of goods to Company C. In the case of a hardware product delivered directly to a client's site, only the model and configuration details are required.

4.13 CONTROL OF NON-CONFORMING PRODUCT

Those actions pertaining to the control of non-conforming products are covered by

QPC 017—Non-conformance, Corrective Action and Records

4.14 CORRECTIVE ACTION

All corrective actions that are necessary on products which do not conform will be captured by:

(a) The review process defined in QPC 006—Software Review.
(b) The audit process defined in QPC 019—Internal Quality Audit using the audit report which, in turn, would generate a system amendment request.
(c) The change control procedure defined in QPC 004—Software Development and QPC 002—Configuration and Change Control.
(d) The error reporting procedure defined in quality procedure QPC 017—Non-conformance, Corrective Action and Records.

The interrelationship can be summarised as shown in Fig. 2.

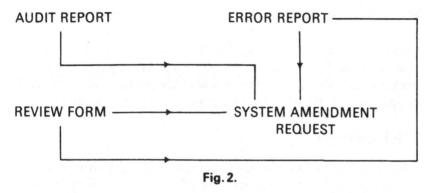

Fig. 2.

Only an uncontrolled product can be changed by an error report form. If the product has been uplifted to a controlled status then a system amendment request must be issued.

4.15 HANDLING, STORAGE, PACKAGING AND DELIVERY

4.15.1 General

All products supplied by Company C will be recorded and a delivery note supplied with the goods. The delivery method will be determined by the product, and could be an authorised carrier (refer to list of approved suppliers) or a member of staff.

4.15.2 Handling

Staff are instructed in the safe handling of items so as to prevent deterioration.

Due to the nature of the products distributed by Company C the handling instructions are:

(a) *Software:* all software is supplied on magnetic media and must be in the appropriate packaging supplied with that media. The media should not be subjected to extremes of temperature or humidity and should be kept away from other strong magnetic sources unless the packaging is screened against magnetic fields.

(b) *Hardware:* the appropriate packing instruction for the hardware should be observed.

(c) *Documents:* all documents should be securely bound and be packaged to avoid damage by the method of delivery selected.

121

4.15.3 Storage

Secure storage areas are provided for all finished products. The on-site storage is in the form of lockers. Off-site storage is provided by means of a bank deposit box which is updated whenever the product version is revised or updated. See QPC 011—Goods Inwards, Inspection and Stores.

4.15.4 Packaging

All products are packaged in the containers that are provided upon receipt at Company C. The software items should be marked with the product name and version number.

4.15.5 Delivery

All hardware products which exceed £10 000 in cost shall be delivered by specialist carriers (refer to list of approved suppliers). Carriers of other products should have adequate insurance to cover the cost to Company C.

4.16 QUALITY RECORDS

Records are maintained which verify the effective operation of the quality system. These records include:

(a) identification report
(b) feasibility study
(c) proposal and contract
(d) system description
(e) test plan
(f) review forms
(g) audit reports
(h) error reports
(i) system amendment requests
(j) purchasing records

Items (a) to (c) are kept in the customer file and will remain on file for 3 years, after which they will be archived for a further 2 years.

Items (d), (e) and (i) are kept in the project file and will remain there for a period of 1 year after the completion of the project, after which they will be archived for a further 2 years.

Items (f) to (j) and a copy of item (i) will be kept in the corresponding review file, audit file and system amendment request file, all of which are held centrally by the technical director. They will remain on file for a period of 1 year, after which they will be archived for a minimum of a further 2 years.

QPC 018—Quality Records

4.17 INTERNAL QUALITY AUDITS

Internal quality audits are carried out to verify that quality activities comply with the planned arrangements and also to determine the effectiveness of the quality system.

Audits are carried out on a regular and planned basis. The results and corrective actions are notified to the board of directors. There shall be a re-audit within a specified period to determine the status of the corrective action.

There is no separate procedure for this item but the technical director will maintain a file of planned and past audits together with details of results and corrective action.

QPC 019—Internal Quality Audit

4.18 TRAINING

4.18.1 Recruiting and Training Programme

Manpower resources and facilities in the company are periodically and systematically reviewed against past, present, planned and forecast levels of business activity by product type, volume and mix to determine and regulate the forward programmes of recruitment and training from which future manpower resources will be met.

The technical director is responsible for specifying the minimum entry requirements, appraisal of applicants, and selection and induction of the new entrant. He is also responsible for identifying and satisfying the training needs of all existing personnel in his areas of operation.

QPS 020—Education and Training

4.19 SERVICING

Contracts could include a requirement for servicing and software maintenance whereby the quality of the product is maintained.

The duties called for within a service contract would be performed by suitably qualified personnel having the required skills and experience.

QPC 021—Servicing

4.20 STATISTICAL TECHNIQUES

Procedures will be established to verify the acceptability of a product and any processes.

These procedures will establish statistical techniques, quality measures and their implementation. The data used in the quality measures will be collated and collected against a product specification in terms of its conformance to its specification and may include

QPC 022—Statistical Techniques

5 LIST OF QUALITY PROCEDURES

Procedure number   *Title*

Procedure number	Title
QPC 001	Contract Review
QPC 002	Configuration and Change Control
QPC 004	Software Development and Standards
QPC 005	Software Design Standards
QPC 006	Software Review
QPS 007	Subcontracted Equipment and Software
QPC 008	Documentation Format and Standard
QPC 009	Purchasing and Suppliers
QPC 011	Goods Inwards, Inspection and Stores
QPC 015	Test
QPC 017	Non-conformance, Corrective Action and Records
QPC 018	Quality Records
QPC 019	Internal Quality Audit
QPS 020	Education and Training
QPC 021	Servicing
QPC 022	Statistical Techniques

6 LIST OF WORK INSTRUCTIONS

To be added, as appropriate, by the company in question.

PART 5

Sample Procedures

COMPANY A

QUALITY PROCEDURE QPA 002

CONFIGURATION AND CHANGE CONTROL

Compiled by...

Checked by...

Approved by...

Date..

CHANGE HISTORY PAGE

Document status	Date issued	Number of pages	Changed pages	Change/ defect no.
XX	XX	XX	XX	XX

CONTENTS LIST

1 INTRODUCTION

1.1 PURPOSE

To ensure item and finished product integrity, a configuration management programme is needed which will ensure continued integrity of items during product development.

1.2 SCOPE

All mechanical, electronic and electrical items and finished products produced or purchased by Company A or as directed by contract constraints.

1.3 TERMS AND ABBREVIATIONS

CCB: Change control board—this reviews, evaluates and authorises changes to CIs at critical points. Currently, at Company A, the CCB consists of the engineering and quality manager, the inspector and the sales manager.

CI: Configuration item—CIs are the smallest uniquely identifiable items which, when combined with other items, provide units and modules of the finished product.

CFA: Configuration functional audit—CFAs are an audit of the CIs against their functional requirement.

CPA: Configuration physical audit—CPAs are 'as built' audits of the CI.

2 APPLICABLE DOCUMENTS AND REFERENCES

QPS 001—Contract Reviews
QPS 003—Design and Review
QPA 011—Goods Inwards and Inspection
QPA 017—Non-conformance, Corrective, Action and Records
QPA 014—Line Inspection and Test
QPA 010—Control of Production
QPS 013—Stores

3 PRODUCTION CONFIGURATION MANAGEMENT

A production configuration management programme will be established. Control points (baselines) will be agreed at critical points in the production life-cycle which need to be controlled.

CIs will be reviewed and evaluated by the engineering and quality manager against the control points, depending on product and contract requirements.

To establish a control point, the CIs to be included need to be agreed; then the monitoring and control mechanism, by which production will be managed, must be agreed.

Reporting by engineering and quality assurance will then take place at regular intervals to establish the status of the CIs against the control points.

3.1 PRODUCTION DATA MANAGEMENT

Data management is concerned with the control and monitoring of all documents, including the contract, assembly specifications, drawings and design sheets, which comprise the production folder.

Data management ensures that the documentation is consistent with the CIs and finished product, across the life-cycle.

3.2 IDENTIFICATION OF CONTROL POINTS

Production configuration management should identify three critical points in the life-cycle. CCB reviews are held at these points and are at the end of:

(a) *Specification* (or requirement). This reviews and agrees the specification and functional requirements which the finished item must satisfy. To include contract and purchasing requirements.

(b) *Production*. This agrees that the finished item satisfies its assembly, drawing, kitting and design specification.

(c) *Product* (finished item). This reviews and confirms after final testing, but prior to release, that the product conforms to all contractual and specification requirements.

3.3 CONTROL

After a control point is established, unauthorised changes to CIs and documents will not be allowed, unless the change has been reviewed and

QPA 002
Issue 1

Date	Project	Project No.	Change No.

Subject/Item:

Raised by:

Details of Change:

Analysis of Change by CCB:

Cost: Schedule:

Documentation: Reliability:

Testing: Safety:

* Agreed CCB: (Signature of reviewers)
* Not agreed CCB: (Signature of reviewers)

(* Delete as applicable)

Date ....

Copy to Engineering and Quality Manager.

Fig. 1. Change request.

evaluated by the CCB. All proposed changes must be documented. See Fig. 1.

The same controls apply as with the concessions/defect sheets. See QPA 017—Non-conformance, Corrective Action and Records.

3.4 LEVELS OF CONTROL

Depending on the degree of change, levels of CCB control can be exercised.

(a) Level 1—major modifications at any control point which effect requirement specifications, safety, reliability, contract and need explicit client agreement.

(b) Level 2—minor modification to CIs or documentation which require agreement with the client.

(c) Level 3—modifications which are internal to Company A and do not need client agreement.

4 CONFIGURATION AUDITS

These will be performed by the engineering and quality manager or his nominated representative.

The FCA is a means of validating that the production of a CI has been completed satisfactorily; it is a formal prerequisite to the PCA. The following list describes the FCA process:

—Review test/analysis results ensure that the testing is adequate, properly performed and, if applicable, certified by the customer.

—Review test/analysis results verify that the actual performance of the CI complies with its specifications and production requirements, and that sufficient test results are available to ensure the CI will perform in its application environment.

—Review all defect, concession and change requests to approved customers' specifications and standards. The purpose is to determine the extent to which the equipment undergoing FCA varies from applicable specifications and standards, and to form a basis for satisfactory compliance therewith.

The PCA is a means of establishing that the product configuration identification used initially for the production and acceptance of CI units satisfies its requirements in the 'as built' state.

Perform the following activities to ensure a successful PCA:

—Compare drawings with equipment to ensure that the latest drawing change numbers have been incorporated into the equipment, that part numbers agree with the drawings, and that the drawings are complete and accurately describe the equipment. Review parts kits to ensure client approval of non-standard parts.

—Verify that acceptance test results have been reviewed to ensure that testing is adequate, properly performed and, if necessary, agreed with the client.

—Verify that shortages and unincorporated design changes are listed in a report and have been reviewed.

—Examine the report to ensure that it defines the equipment adequately and that unaccomplished tasks are included as defects/deficiencies.

—List all defects/concessions/changes to approved customer specific-ations and standards. This forms the basis for satisfactory compliance with customers' specifications and standards.

—Review the production release and change control procedures to ensure that they are adequate to control the processing and formal release of drawing changes.

The production folder system is crucial to configuration management. The folder and its contents must always be controlled, be up to date, and be kept securely when not in use.

The production folder will be audited regularly by the engineering and quality manager to monitor and control its contents against specifications, kit and parts lists, drawings, etc.

Reports on FCA and PCA results will be made periodically by the engineering and quality manager.

COMPANY A
QUALITY PROCEDURE QPA 008
DOCUMENTATION FORMAT AND STANDARD

Compiled by...

Checked by...

Approved by...

Date...

CHANGE HISTORY PAGE

Document status	*Date issued*	*Number of pages*	*Changed pages*	*Change/ defect no.*
XX	XX	XX	XX	XX

CONTENTS LIST

1 INTRODUCTION

1.1 PURPOSE

This procedure is intended to provide the detail needed for the format, layout and identification of quality-related documentation; also the issue control of such documentation.

1.2 SCOPE

The scope of this procedure is quality documentation produced by Company A.

1.3 TERMS AND ABBREVIATIONS

None.

2 APPLICABLE DOCUMENTS AND REFERENCES

QPA 002—Configuration and Change Control

3 TEXT LAYOUT

To aid both ease of production and updating of documents, the traditional method of decimalised topic headings will be used. The number of levels of headings, and subheadings, should be limited to aid readability. The first level of heading should be in upper-case. It should have a left margin sufficient to allow for binding. The second level should be in lower-case except for the initial letter. An example overall layout is:

9.0 HEADING

9.1 Subheading

Where extensive detail has to be listed at any of the levels, then lower-case alphabetic labelling will be used for each item listed. This more detailed layout is:

9.0 HEADING

(a) Quality topic A

(b) Quality topics B and C

9.1 Subheading
(a) Quality topic A detailed
(b) Quality topic B detailed
(c) Quality topic C detailed

4 FORMAT

4.1 PAGE NUMBERING AND IDENTIFICATION

All pages of quality documents will be numbered sequentially, starting at 1, page by page. If any section of a document is sufficiently large the page-by-page sequential number may be prefixed by the section number throughout this section.

All page numbers should be placed at the top right-hand corner or bottom centre of the page. All pages will carry the document identity and issue in the top right-hand corner.

4.2 TITLE PAGE

All quality-related documents produced within Company A will carry a standard front page (see Fig. 1). Quality documents which are general in application need not carry any reference to 'Prepared for' or 'Contract no.' unless specifically tailored for a project/contract.

All documents will be identified by a unique identifier formed from two constituent parts, i.e. head and tail. These are discussed fully later.

Details on document status, issue date and review (compiled, checked, approved) are also discussed later.

4.3 CHANGE HISTORY PAGE

Page 2 of all quality documents will be a change history page, recording the document's status and its issue dates, the number of pages of that issue, changed pages and change/defect numbers against which changes are made (see Fig. 2).

4.4 CONTENTS PAGE

Page 3 of all quality documents will be the contents page. It will list the two or three levels of headers and their associated numerical identifiers. Where

Document Identifier : head and tail
Document Issue Status : alpha/numeric

COMPANY A

Quality Document Type and Number

QUALITY DOCUMENT TITLE

Prepared for:

Contract:

Compiled by:

Checked by:

Approved by:

Date approved:

Fig. 1. Quality related document front page.

Document status	Date issued	Number of pages	Changed pages	Change/defect number
A	04/01/88	12	N/A	N/A
B	20/02/88	16	N/A	N/A
1	01/04/88	16	13–16	37
2	11/08/88	16	9, 12, 14	49

Fig. 2. Typical change history page.

the contents are sufficiently brief the change history and contents pages may be accommodated on the same page.

4.5 CONTENTS SPACING

Documents will be typed in single-line spacing, except where contract requirements define otherwise.

4.6 CONTENTS

As appropriate, all quality documents will follow the layout as detailed below.

(a) *Introduction (Section 1):* each document shall carry a brief introduction giving the purpose and scope of the document, and any special terms and assumptions.

(b) *Applicable documents and references (Section 2):* all applicable documents and references should be listed. Where these are external to Company A details should also include author's name, publishers and date of publication.

(c) *Text layout of document (Section 3 onwards):* main body of document, following the text layout requirements as detailed in Section 3.

(d) *Figures:* figures should be collected together, possibly at the end of the document, and referred to by number and title.

(e) *Appendices:* appendices will be treated like sections, but identified by the letters A, B, C, etc., and title.

5 DOCUMENT IDENTIFICATION

All quality documents produced by Company A (except where a contract specifies otherwise) will carry a unique identifier. The identifier will be formed from two constituent parts, a head and a tail.

5.1 HEAD AND TAIL IDENTITY

The head will be derived from the quality document type, i.e. QM for quality manual and QP for quality procedure.

The tail will be a simple serial number starting at 001, allocated by the engineering and quality assurance manager within each type. If quality documentation has to be specifically prepared for a project/contract, the identifier will have a third constituent part added; it will comprise the acronym of the project/contract.

5.2 DOCUMENT STATUS AND ISSUE CONTROL

The status and issue of a quality document will be shown, like its identifier, on the top right-hand side of the title page.

5.2.1 Status Control

There are three types of status control:

(a) *Uncontrolled:* here documents may be changed easily by the compiler/checker without further reference.
(b) *Controlled:* here documents may be changed only after a review and approval. These changes are internal to Company A.
(c) *Frozen:* change here involves both internal review and agreement with the client.

5.2.2 Issue Control

There are two types of issue control:

(a) *Alphabetic issue:* these documents are not subject to review and can be changed without further reference by the compiler/checker. Alphabetic issue is related to uncontrolled status control.
(b) *Numeric issue:* these documents are subject to review and can only be changed after the appropriate review. This issue type is related to frozen and controlled status control.

5.3 CONTROLLING CHANGE

Quality documents with a controlled or frozen status shall only be updated after a change/defect form is raised and reviewed, and the update agreed at the review. For full details see QPA 002—Configuration and Change Control.

When an update is made to a numeric issue controlled quality document, the issue status of the title page will be incremented by one, the appropriate signatures obtained, and the date changed to reflect the increment. The change history page will also be updated.

Copies of controlled and frozen documents will only be made with the explicit agreement of the company engineering and quality manager. Then distribution/circulation will be controlled/limited.

6 OTHER QUALITY DOCUMENTATION

Work instructions will be created for specific purposes. These will be less complex documents than the procedures. They will contain the standard format (except for contents page).

COMPANY A
QUALITY PROCEDURE QPA 010
CONTROL OF PRODUCTION

Compiled by..

Checked by..

Approved by..

Date..

CHANGE HISTORY PAGE

Document status	Date issued	Number of pages	Changed pages	Change/ defect no.
XX	XX	XX	XX	XX

CONTENTS LIST

1 INTRODUCTION

1.1 PURPOSE

To provide a procedure to control materials throughout production and into stores and despatch.

1.2 SCOPE

Items manufactured at Company A.

1.3 TERMS AND ABBREVIATIONS

ISO: internal sales order

2 APPLICABLE DOCUMENTS AND REFERENCES

QPS 001—Contract Review
QPS 009—Purchasing and Suppliers
QPS 013—Stores

3 ADMINISTRATION (Production)

The administration department is responsible for the following functions:

—entering and processing of customers' orders
—raising purchase requisitions on outside suppliers
—control of factory stores
—despatch and invoicing

which are detailed below.

4 ENTERING AND PROCESSING OF A CUSTOMER'S ORDER

On sales receiving a customer's order (this may include a telephone order) a design sheet will be prepared by the sales department. Administration will:

(a) Allocate an internal sales order no. (ISO; see Fig. 1). This is done automatically by the computer.

Acc. No.		O/No.				Date:	Invoice Address	I.S.O.:	Despatch Address
Insp. ———	Replace ———	Stack ———		Shipping ———					

ASSY CODE:

Item No.	Product description	Quantity	Date required	Date promised	Unit Price £/$
01					
02					
03					
04					
05					
06					
07					
08					
09					
10					

Items per Assembly

Special Instructions:
New Drawing : Yes/No
Customers Drawing :

Label Typing Instructions : Quantity : ....

Fig. 1. Internal sales order.

(b) Enter details of the customer's order into the computer, making sure that any special requirements are identified. In the case of special inspection or certification an appropriate entry is made in the special requirement box of the design sheet and the original of the order is sent to engineering and quality (inspection). Administration should file a copy of the order.

(c) After the internal sales order has been generated, the contract detail is entered in the administration log book.

(d) Generate a production and acknowledgement copy of the ISO. A copy of the ISO will be placed in the production folder (made by administration) together with the design sheet.

(e) These are issued, by administration, to stores in the production folder.

(f) Send the acknowledgement copy of the ISO with the delivery promise to the customer. The delivery date is estimated by a pre-kitting exercise, consulting a relevant copy of the latest available parts list.

5 STORES (Kitting)

On receiving the production folder from administration, the storeman will check availability and prepare a parts list. One copy of the parts list will be added to the production folder. If components are required to be manufactured 'in-house', the storekeeper will make out a works order (Fig. 2) giving the part number and quantity required and passing the production folder to the workshop.

On completion the workshop will return the parts list with the parts and retain a copy for costing. 'Bought out' components are obtained by the storeman raising a purchase requisition and passing it to administration. If approved components are required, a certificate of conformity should be requested.

Having accumulated all the required components, the storeman will amend the stores records and pass all the components with the production folder to the production department. The storekeeper will retain a copy of the parts list for his own records.

6 PURCHASE ORDERS

On receiving a purchase requisition from stores, purchasing will place an order on an approved supplier (a list of approved suppliers is held in the

WORKS ORDER

TIME RECORD

Week ending	INITIALS	Mon.	Tues.	Wed.	Thur.	Fri.	Sat.	TOTAL

Commenced / /   Completed / /   Total

Finished Goods Note

Date

Quantity

MATERIALS USED

Quantity	Code/Description	£	p

TOTAL MATERIAL COST

LABOUR

TOTAL WORKS COST

UNIT COST

DESIGN OFFICE INSTRUCTIONS
1. Prepare production drawings as per attached D.I. sheet
2. Prepare drawings as follows:-

WORKSHOP INSTRUCTIONS
1. Manufacture piece-parts as stated.
2. Manufacture the following:-

REQUESTED BY:-	DEPT.		DATE / /	COST CENTRE	Req'd. Wk. ......
DRG./Code No.	QUANTITY	DELIVERY	DESCRIPTION	ISSUED	W/O No.
		Wk.No.		/ /	

Fig. 2. Works order.

146

administration department). If approved components are required the component specification should be quoted and a certificate of conformity requested. Copies are filed by administration and goods inwards.

7 CONTROL OF FACTORY STORES

A record of the stores contents is held in stores and a duplicate copy is maintained by the administration department. This information is updated by virtue of the continuous computer entries. For components which are in continual demand, a minimum stock level will be entered on the computer. When this level is reached an automatic reordering sequence is invoked on the computer-issued records.

8 PRODUCTION

On completion of the assemblies, the supervisor will pass the design sheet to administration for the creation of equipment labels. These are returned for attaching to the assemblies together with the design sheet.

9 DESPATCH AND INVOICING

After the production department has completed the job and inspection has passed the assembly (or assemblies), the inspector will stamp the design sheet and the ISO and pass them to administration, who will raise despatch notes and, if required, a certificate of conformity.

These documents are passed to despatch, who ship the goods to the customer, making sure the packaging is robust enough to prevent damage during transit (WI XXX).

A copy of the despatch note is added to the production folder and returned to administration together with the design sheet, routing slip and any other appropriate documents.

The invoice is sent to the customer. Copies of the order, ISO and invoices are then filed by administration in the production folder and all forwarded to the engineering and quality department as part of the quality records.

COMPANY A

QUALITY PROCEDURE QPA 011

GOODS INWARDS AND INSPECTION

Compiled by...

Checked by..

Approved by ....

Date..

CHANGE HISTORY PAGE

Document status	Date issued	Number of pages	Changed pages	Change/ defect no.
XX	XX	XX	XX	XX

CONTENTS LIST

1 INTRODUCTION

1.1 PURPOSE

To describe the procedure for all goods received at Company A.

1.2 SCOPE

All items which arrive at Company A.

1.3 TERMS AND ABBREVIATIONS

None.

2 APPLICABLE DOCUMENTS AND REFERENCES

QPS 022—Statistical Techniques
QPS 012—Identification and Traceability
QPA 017—Non-conformance, Corrective Action and Records
QPS 009—Purchasing and Suppliers
BS 6001—Sampling Procedures and Tables for Sampling by Attribute

3 GOODS INWARDS PROCEDURE

Each purchase requisition has a unique identifier. All items received at goods inwards will be checked against the purchase requisition for quantity and description. Bulk manufactured parts and electrical components will be either 100% inspected or sampled in accordance with the appropriate BS 6001 sampling procedure (see QPS 022—Statistical Techniques). The decision to use sampling techniques will be taken by the engineering and quality manager.

All finished items (assemblies and subassemblies) will be 100% inspected in accordance with the appropriate drawings, specifications or work instructions.

Certificates of conformity will be checked.

3.1 ITEM IDENTIFICATION

Items will be identified as specified in QPS 012—Identification and Traceability.

3.2 NON-CONFORMING ITEMS

All non-conforming items will be segregated at goods inwards and will be under the control of the engineering and quality manager.

The supplier will be contacted by telephone and the problem explained. In the event that the non-conformance is not resolved then the engineering and quality manager will write to the vendor and a copy of the correspondence kept in the vendor file awaiting satisfactory resolution.

COMPANY A

QUALITY PROCEDURE QPA 014

LINE INSPECTION AND TEST

Compiled by..

Checked by..

Approved by..

Date...

CHANGE HISTORY PAGE

Document status	Date issued	Number of pages	Changed pages	Change/ defect no.
XX	XX	XX	XX	XX

CONTENTS LIST

1 INTRODUCTION

1.1 PURPOSE

To describe the inspection procedures needed to control the manufacturing and production process, including the testing of in-house products.

1.2 SCOPE

All those processes involved with on-line and final inspection and test.

1.3 TERMS AND ABBREVIATIONS

ISO—internal sales order
C of C—certificate of conformity

2 APPLICABLE DOCUMENTS

BS 5750 (Part 1) 1987
Customer's specification
Appropriate drawings
QPS 022—Statistical Techniques
QPS 016—Calibration
QPS 001—Contract Review
QPA 010—Control of Production

3 PROCEDURE

3.1 PIECE PART INSPECTION

All parts produced by the production department are to be inspected in accordance with the latest issue of the relevant drawing.

Initially, the first-off part is to be examined for conformity to all aspects of the drawing and, if satisfactory, stamped accordingly. Any discrepancies are to be corrected and verified by re-inspection before quantity production commences.

In-line inspection of the production items will be carried out as indicated, using the inspection instruction by the inspector.

It is the responsibility of the operator to know the standards required and to report any discrepancies or degradation, of which he is aware, that occurs during production.

On completion of a batch the parts are re-inspected to the appropriate drawings and signed off if appropriate.

3.2 ASSEMBLY INSPECTION

For small quantities (up to five assemblies) in-line inspection will not be necessary unless the assembly supervisor identifies a problem. In such a case the engineering and quality department will be consulted and inspection applied as appropriate.

For larger batches assemblies will be mechanically checked at random (QPS 022) using the appropriate drawings.

3.3 FINAL TEST

All assemblies involving semiconductor components will be tested at low voltage using the assembly test equipment provided.

Requirements derived from the customer's specification will also be tested.

In the case of assemblies having only a mechanical function (e.g. clamps) sample inspection will be adequate and test will not apply unless specifically called for.

4 CONTRACT COMPLETION

On completion of inspection and test, the contract routing slip and the design/test sheet are to be signed by inspection. These will be returned to administration, in the production folder, as notification of readiness for despatch.

An entry is to be made in the inspection day book, noting the contract number, assembly number, and quantity, customer and date.

4.1 CERTIFICATE OF CONFORMITY

When an order specifies a C of C it is the inspector's responsibility, when the order is ready for despatch, to raise a certificate as follows:

(a) Obtain ISO number and withdraw the customer's order from the production folder.

155

Date:

Contract number:

CERTIFICATE OF CONFORMITY/CONFORMANCE

Batch of ....

Part No. ....

ISO No. ....

Invoice No. ....

Date ....

We have manufactured and inspected in accordance with our drawing/
specification.

Signed .... Engineering and Quality Manager
(or his authorised deputy)

For and on behalf of company A.

Fig. 1. Certificate of conformity/conformance.

(b) Check the order details to ensure that the terms have been fulfilled.
(c) Enter the contract order number on the C of C together with the other information called for on the certificate (see Fig. 1).
(d) Stamp and sign the C of C.
(e) Send copies:
  —with goods
  —to customer's quality manager
  —to Company A record file

COMPANY A

QUALITY PROCEDURE QPA 017

NON-CONFORMANCE, CORRECTIVE ACTION AND RECORDS

Compiled by..

Checked by..

Approved by..

Date..

CHANGE HISTORY PAGE

Document status	*Date issued*	*Number of pages*	*Changed pages*	*Change/ defect no.*
XX	XX	XX	XX	XX

CONTENTS LIST

1 INTRODUCTION

1.1 PURPOSE

The purpose of this procedure is to control non-conforming items, either when first received at goods inwards or during assembly and testing of the manufactured article. A closed-off and controlled area will be provided for the segregation of non-conforming items.

1.2 SCOPE

All items, from goods inwards to the manufactured item ready for despatch, are subject to this procedure wherever a non-conformance is found.

1.3 TERMS AND ABBREVIATIONS

None.

2 APPLICABLE DOCUMENTS AND REFERENCES

QPS 016—Calibration of Measuring Equipment
QPS 009—Purchasing and Suppliers
QPA 002—Configuration and Change Control
QPS 003—Hardware Design and Review
QPS 011—Goods Inwards and Inspection
QPA 014—Line Inspection and Test
QPA 010—Control of Production
QPS 013—Stores

3 CONCESSION/DEFECT REPORTING

If a controlled item (one under configuration control) is found to be defective in any way, a concession or defect sheet will be raised. See Fig. 1.

Sheets will be assigned a unique serial number for identification purposes. A register will be kept by the engineering and quality manager, who will allocate the unique number.

Date   Project   Project No.   Sheet No.

Subject/Item:

Raised by: Date:

Details of defect: Change request number:

Copy to Engineering and Quality Manager:

Corrective actions:

Analysis of defect:

Cost: Schedule:
Documentation: Reliability:
Testing: Safety:

Defect correctable
Not correctable, concession needed:
Agreed Engineering and Quality Manager:

Signed ....Date ....

Fig. 1. Concession/defect sheet.

4 CORRECTIVE ACTIONS

On discovering a defect, the engineering and quality manager must be informed. He is responsible for initiating corrective actions and seeing that the sheet is properly completed. Where necessary the sheet will be supported by appropriate documentation.

Corrective actions may involve:

(1) correspondence with supplier
(2) correspondence with client in the case of client-issued equipment

160

(3) report of in-house work/rework at any stage of production
(4) concession/defect sheets
(5) change requests (see QPA 002; Fig. 1)
(6) scrap decisions

Any in-line and production documentation must be suitably amended to indicate the status of the item. Any item in stores and purchasing will be treated similarly. All non-conforming items must be segregated and be under the control of the engineering and quality manager.

Copies of sheets must be included in the production folder.

5 SENTENCING

If it is found that a defect cannot be corrected, a concession must be raised. If it can be corrected, a defect will be noted and action taken to clear the defect. If a change request has been raised as a result of a defect, its number must be added to the sheet. The appropriate concession/defect sections of Fig. 1 will be completed.

The concession will acknowledge that a deviation from the requirements has been found and that, where appropriate, there will be a work-around or scrap. This must be agreed by the engineering and quality manager. With a defect, the defect will be corrected and the engineering and quality manager's agreement obtained.

6 DEFECT ANALYSIS AND REVIEW

A monthly summary of the preceding month's reports will be prepared by the engineering and quality manager.

Where specific trends become apparent, the engineering and quality manager will investigate further. This may involve contacting a supplier or seeking the approval of the managing director to change suppliers or change in-house practices and procedures.

COMPANY A

QUALITY PROCEDURE QPA 019

INTERNAL QUALITY AUDIT

Compiled by...

Checked by...

Approved by...

Date...

CHANGE HISTORY PAGE

Document status	Date issued	Number of pages	Changed pages	Change/ defect no.
XX	XX	XX	XX	XX

CONTENTS LIST

1 INTRODUCTION

1.1 PURPOSE

To ensure that the Company A quality system is audited against the current documented standards and procedures.

1.2 SCOPE

All aspects of the Company A quality management system.

1.3 TERMS AND ABBREVIATIONS

None.

2 APPLICABLE DOCUMENTS AND REFERENCES

All documents listed in Quality Manual QMS 001.

3 PROCEDURE FOR AUDITS

3.1 POLICY

Apart from periodic external assessments by companies and authorities, each section within Company A is subjected to audits every 12 months by the engineering and quality manager or his nominated representative (auditor). An annual audit chart is displayed in the quality section which identifies the respective review dates.

The audit is carried out to check that the procedures, work instructions and systems laid down in the company quality manual are being adhered to, and to check that these remain adequate to ensure the required standards.

Between audits it is the responsibility of the engineering and quality manager to ensure that the procedures, work instructions and systems are adhered to.

The auditor shall carry out audits as described below and institute corrective actions where necessary, including modification, addition or replacement of the procedure or work instruction as the circumstances may indicate. The auditor will record the date and result of the audit, and any corrective action necessary. The results of any re-examination following such corrective action will also be recorded.

3.2 METHOD

In each area the auditor will formally communicate with the manager and suggest a convenient date for the audit.

The auditor will require confirmation that there have been no changes in procedures, processes or facilities that could necessitate a procedural review, and that no specific problems have arisen in connection with orders that might require investigation or procedural/process changes.

When visiting an area the auditor will liaise with the manager and decide with whom that audit will be conducted.

Conformance in each area of managerial responsibility will be established by a general appraisal of the area with random spot checks pursued on specific items to establish continuity of conformance.

The extent of the audit will be controlled by the quality procedures to be covered, rather than by the amount of time available.

If all sections have not been completed in a reasonable time, then a further visit is to be arranged to complete the programme.

The results of the completed audit will be entered on the internal quality audit form and referred to the managing director and the engineering and quality manager for agreement and authorisation (see Fig. 1).

4 CORRECTIVE ACTION

Any discrepancies found during the quality audit shall be agreed by the auditor and the area concerned, and recorded on a corrective action form (see Fig. 2). A copy of the corrective action form shall be retained and signed for by the engineering and quality manager and the managing director.

5 CORRECTIVE ACTION FOLLOW-UP

On the scheduled date for completion of the corrective actions, the engineering and quality manager or his representative shall ensure the successful completion of each corrective action itemised on the corrective action form.

Internal QUALITY AUDIT		
Department/Area audited:	Date of audit:	Report No.: QA.....................
Basis of audit:		
Result/Non-conformance: Signed .... (Manager)		
.... (Date) .... (Auditor) .... (Date) (Managing Director) .... (Date) (E. & Q. Manager) Next audit due		

Fig. 1. Internal quality audit.

166

Internal QUALITY AUDIT Corrective Action		
Department/Area audited:	Date of audit:	Report No.. QA.....................
Corrective action:		

Scheduled date of completion: Signed: ....
(Manager)

Action pending: ....
(Date)   (Auditor)

Action: Followed up: ....
(Date)   (Managing Director)

Action: Complete: ....
(Date)   (E. & Q. Manager)

Fig. 2. Internal quality audit, corrective action.

If the corrective actions are still outstanding then a new completion date shall be agreed by the auditor and the managing director. A signed follow-up copy of the corrective action form shall be retained by the managing director and the engineering and quality manager.

The above corrective action follow-up will apply until the outstanding deficiencies have been completed.

6 AUDIT DISAGREEMENTS

Disagreements between the managing director and the quality management which occur as a direct result of the findings of the internal audit shall be the subject of a meeting of all parties concerned, and the directors, who will endeavour to bring the dispute to an acceptable conclusion.

COMPANY A OR B
QUALITY PROCEDURE QPS 001
CONTRACT REVIEW

Compiled by...

Checked by...

Approved by ....

Date...

Note: Reference is made to the engineering and quality manager, which is specific to Company A. The appropriate title will be used according to your company.

CHANGE HISTORY PAGE

Document status	*Date issued*	*Number of pages*	*Changed pages*	*Change/ defect no.*
XX	XX	XX	XX	XX

CONTENTS LIST

1 INTRODUCTION

1.1 PURPOSE OF DOCUMENT

All contracts and orders will be reviewed and checked for correctness before and after acceptance. During the life of the order and contract, periodic reviews and coordination of activities shall take place to ensure continued contract conformance.

1.2 SCOPE

Any contract or order contemplated by Company A or B will be handled in accordance with this procedure.

1.3 TERMS AND ABBREVIATIONS

None.

2 APPLICABLE DOCUMENTS AND REFERENCES

QPA/B 002—Configuration and Change Control
QPS 009—Purchasing and Suppliers
QPA/B 010—Control of Production
QPA/B 011—Goods Inwards and Inspection
QPA/B 017—Non-conformance, Corrective Action and Records
QPS 007—Subcontract Equipment and Software

3 CONTRACT ADMINISTRATION

A contract is an agreement, verbal or written, to perform a service or provide a product, at some future time, for a financial consideration. Companies A or B can be either the supplier or the customer in a contract. The principles of this procedure apply equally to either.

Contract administration is concerned with the management of the contract from its initial agreement through to completion. It is also concerned with any legal ramifications. It also ensures that all activities are handled in accordance with the contract.

3.1 PRE-CONTRACT ACCEPTANCE REVIEW

Before a contract is accepted by either Company A or B, a detailed review of the contract shall have taken place.

For standard contracts, this will be against a checklist of items agreed by the sales and the engineering and quality managers. For non-standard contracts, the above managers will prepare a list of conditions with an explanation of the implications of each.

These non-standard conditions will be thoroughly reviewed against Company A's or B's capabilities to satisfy the condition. Non-standard conditions could impact on:

—cost and time
—penalty clauses
—security requirements
—resources (staff, software, hardware)
—validity of proposal and estimate
—acceptance criteria
—customer-imposed quality conditions
—schedule of deliverable items
—subcontractors

3.2 CONTRACT ACCEPTANCE

All contracts and orders from customers will be scrutinised, at acceptance, by the sales manager to confirm that the conditions of the contract can be met. If any conditions are still at variance with those of Company A or B then the sales manager will negotiate with the customer to agree the conditions.

The engineering and quality manager will be involved in these negotiations as necessary.

All contracts must be confirmed in writing, facsimile or telex. Until confirmation is received, the contract will not be processed. Only the managing director's written authority will enable work to start.

3.3 CONTRACT MONITORING

During the project, the project manager will regularly brief the sales and the engineering and quality managers with details of contract or order progress, both technically and financially.

172

The briefing will include such topics as:

—Confirming that any changes to the requirements have been agreed and that appropriate change request and defect/concession sheets are signed off.
—That for any change or defect/concession an analysis of schedule and cost implications has been agreed.
—That non-standard conditions are still being satisfied (e.g. special inspection/test conditions).
—Changes in customer organisation and any impact on the customer's financial, technical and quality responsibilities.
—Demonstrations to the customer of the developing product.
—Distribution of correspondence to all Company A or B managers of letters, facsimiles, telexes, etc., between the company and the customer, in order to ensure that the management are fully aware of progress and problems.

3.4 CONTRACT AUDITS

With the signing of the contract, a legal commitment has been made to produce and deliver a system/product for the customer.

The contract is the first document of the project. It needs to be controlled and kept under strict configuration and change control. All records which impact on the contract need to be regularly checked, reviewed, agreed and stored.

The engineering and quality manager will make periodic internal audits of conformance to the contract. The frequency of such audits will vary with the technical and financial complexity of the contract.

Depending on the complexity of the contract, the customer may request to have his own auditor/assessor carry out these audits.

COMPANY A OR B
QUALITY PROCEDURE QPS 003
DESIGN AND REVIEW

Compiled by...

Checked by..

Approved by...

Date...

CHANGE HISTORY PAGE

Document status	*Date issued*	*Number of pages*	*Changed pages*	*Change/ defect no.*
XX	XX	XX	XX	XX

CONTENTS LIST

1 INTRODUCTION

1.1 PURPOSE

Hardware designs must be adequately reviewed to ensure completeness and consistency. To ensure that these items are verified, peer reviews are conducted.

Formal reviews must be structured to demonstrate the hardware design's conformance to requirements and specifications, and its acceptability to the client.

1.2 SCOPE

The scope of the peer reviews and formal reviews will vary from project to project. They may therefore be combined or reduced in scope, depending on contract circumstances.

1.3 TERMS AND ABBREVIATIONS

CI—configuration item: the smallest uniquely identifiable item, which when combined gives units, modules, components, etc., of the finished product.

2 APPLICABLE DOCUMENTS AND REFERENCES

QPA/B 002—Configuration and Change Control
QPB 004—Software Development and Standards
QPA/B 008—Documentation Format and Standard
WIS XXX—Engineering Drawing

3 HARDWARE DEVELOPMENT REVIEWS

3.1 Reviewing and Inspecting Hardware Development Products

All reviews, inspections and walkthroughs are conducted with the authority of the engineering and quality manager.

Peer reviews are conducted during the requirements definition, the preliminary design, and during the detailed design. These reviews are useful in revealing problems that may not be seen before demonstrating the product to the client.

Figure 1 identifies the peer reviews that are conducted in each phase of the hardware development life-cycle.

Depending on exact project circumstances, peer reviews and formal reviews can be combined. Any of the following methods and checklists can be chosen as considered applicable by the project or engineering and quality manager.

3.1.1 Peer Reviews

There are two types of peer reviews, known as inspections and walkthroughs.

Peer inspections are normally one-to-one inspections between the hardware engineer working on the configuration item (CI) design and another engineer who may or may not be working on the design. Before a peer inspection is performed, a checklist of the items to be inspected will be provided to the engineer performing the inspection.

Peer inspections can be used to verify compliance with the requirements specification, to develop an understanding of what the component entails, and to verify consistency from one review to the next.

Before an inspection, the inspection engineer is given a synopsis of the design, the requirements for the design and a checklist of items to be inspected. This information should be provided in advance to allow the engineer time to do a thorough inspection and to understand the design.

After completion of the inspection, the inspection engineer forwards the comments to the engineer who designed the item. If all of the requirements have not been considered, then the design engineer must make modifications so as to include them. The inspection engineer has the final word on when the inspection is satisfactorily completed.

Peer walkthroughs, on the other hand, are normally rigorous reviews of the CI components and its associated parts. They are normally conducted by a group of individuals including members of the design team. The primary purpose is to check details of the design. The walkthrough should be used to review a detailed item such as a board design, packaging, schematics or mechanical assembly, or a coded software module. Only one peer walkthrough is performed during hardware development.

The design engineer must ensure that all key personnel are given a copy of the design information appropriate to their expertise. This might provide firmware and register definitions for the firmware engineer; schematics for

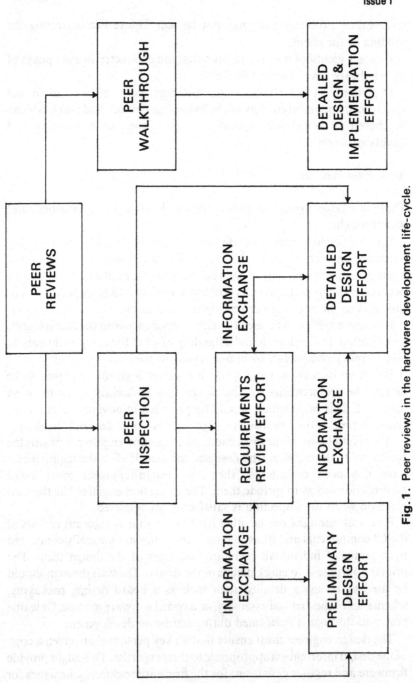

Fig. 1. Peer reviews in the hardware development life-cycle.

178

the hardware engineer reviewing the circuit aspects of the design; board layout and component layout for the engineer performing mechanical design, etc.

The review arrangements and schedule will vary with the complexity of the item. In addition to the information provided to each member of the walkthrough team, a checklist of the relevant hardware requirements and a synopsis of the design will also be provided.

The designer or other engineer will conduct the meeting, which need not cover all topics at once but may be subdivided to avoid confusion.

After completion of the walkthrough, comments are provided by each member of the walkthrough team to each presenting engineer. These comments are reviewed and, if deemed appropriate, addressed to resolve any discrepancies. It is the responsibility of the design engineer to approve changes to a design after the walkthrough process is completed.

3.1.2 Reviews during the CI Requirements Definition Effort

The primary purpose is to ensure that the CI's detailed requirements are consistent with the system requirements and to verify that the derived requirements are valid. The functional design concept and all derived requirements should be subject to peer inspections.

These peer inspections should:

—Verify consistency with system requirements.
—Validate the rationale for establishing each requirement.
—Review the results of supporting studies and analysis.
—Verify the impacts and implications of the requirements.
—Carefully review all derived hardware requirements. The following is a description of some of the most common requirements categories.

—*Interfaces:* ensure that all external interfaces have been identified and their requirements specified to the greatest extent possible. Verify the following:
  ● logical requirements such as communications protocols and standards;
  ● physical requirements such as the type of connectors and type of cabling;
  ● electrical requirements such as voltage levels and data formats.
—*Physical considerations:* determine if component-specific physical characteristics have been considered as follows:
  ● commonality of hardware elements within the CI with other CIs and with existing hardware items;

179

- physical requirements such as rack dimensions; mobility and transportability; weight and floor loading; heat, ventilation and cooling; and facility power.
—*Operation and maintenance:* verify that operation and maintenance requirements are considered. Review factors such as:
- diagnostics and self-tests;
- equipment access;
- interchangeability and commonality of components;
- troubleshooting and repairs;
- logistics;
- reliability, maintainability and availability: verify that requirements are consistent with the system requirements; identify the need for redundant elements to eliminate single points of failure;
- environmental considerations: ensure that ambient and operational environmental requirements have been specified; consider temperature, humidity and electromagnetic factors.

3.1.3 Reviews during the Preliminary Design Effort

The aim of this peer review is to identify and correct design deficiencies early in the preliminary design process. These inspections are conducted before publishing formal design specifications and, if possible, before client review.

By this time the following actions have occurred:

—The CI requirements specifications have been agreed.
—The preliminary design specification is complete in draft form.
—The CI functional capabilities are documented to the block diagram level.

Peer inspection of the elements of the functional design is now conducted. Concentrate on the following aspects of the preliminary design and verify that it is:

—traceable to the requirements;
—consistent with its requirements;
—consistent with its related functional elements;
—understandable;
—consistent with the overall system configuration;
—implementable within technical, resource and schedule constraints.

180

Items to be reviewed include the following:

—functional hardware block diagrams;
—firmware flow diagrams;
—input/output (I/O) descriptions (signal and power);
—operational concepts;
—test concepts resource requirements and scenarios.

3.1.4 Reviews/Walkthroughs during the Detailed Design/Implementation Effort

The purpose of this review/walkthrough effort is to validate the detailed hardware design. These reviews are performed before publishing the CI detailed design specification.

The block diagrams used in the detailed design effort are refined versions of those utilised in the preliminary design effort. By this stage the following has occurred:

—All electronic designs are defined.
—Schematic diagrams are available but may not be complete.
—All major material components have been designed or identified.
—Equipment layouts are defined.
—All interfaces have been described.
—Test and build plans have been formulated and documented.

During detailed design/implementation, peer inspections and walk-throughs are carried out to verify the operation of the CI component and circuit boards.

Peer inspection of the critical design/implementation is more of a functional review of the CI component and is not merely for determining whether or not a circuit board is operational. The design, at the circuit board level, is reviewed to determine if all the functions to be performed by that board have actually been addressed and achieved. A partial list of items to be considered is as follows:

—Review the functional power requirements of the components to verify that the component either sources or receives the proper power to/from its interfaces.
—Review the functional I/O to verify that all signals have been considered.
—Review the firmware flow to determine structured design.

181

—Review the CI component memory size to determine whether enough memory has been allocated for the component to meet its requirement.
—Review the test points to verify that all have been considered.
—Verify proper communications paths have been considered.
—Review the component cabinet and card file layout to verify they agree with the requirement specification.

The peer walkthrough of the critical design/implementation is a rigorous review of the CI component and circuit boards. Every aspect of the detailed design is reviewed and the following is a list of items to be considered:

—Review circuit boards, not only to verify that the functions of the boards have been implemented but also to look at timing diagrams to verify operation of synchronous or asynchronous elements.
—Check connectors and connector type to determine proper connectivity with external and internal interfaces.
—Review analogue designs to determine whether the designs are truly operational and whether they have any potential adverse effect on the digital signals that may be on the same circuit board.
—Check the component cooling system used to determine whether there is a potential overheating problem on a board.
—Look at the power supplies to verify there is no potential for overload based on the proposed design.
—Check for potential cross-talk problems in the cabling with respect to differential signals being sourced or received.
—Look at the hardware architecture bus structure to determine if it is being utilised properly.
—Verify that there is a write-up for each circuit board or function, detailing the operation of the board and the functions it will perform.

3.2 FORMAL HARDWARE REVIEWS

All new hardware items and any design changes to the product's baseline are subject to a series of reviews, including a requirements review, a preliminary design review and a critical design review. For small projects or modifications, the review need not be so formal. However, for significant projects, reviews should be formally conducted. An example of a hardware design review form is shown in Fig. 2.

Depending on the nature of the project, the feasibility of combining these reviews with the parallel software reviews should be considered. Reviewing

An inspection/review of the following hardware item has been performed.

Item:

Time & Date:

Participants:

Engineer:

Team Leader:

Production:

Engineering and Quality Manager:

Others:

Inspection/Review Results:

Certified by: Date:

Fig. 2 Hardware design review form.

both the hardware and software aspects of development will provide a better overall picture of the product.

The purpose of these reviews is to demonstrate compliance with requirements and specifications, and, if applicable, to obtain acceptance of intermediate design and development products from the customer. A summary of the formal reviews and their relationship to one another is given in Fig. 3.

Formal Hardware Review	CI Life Cycle Phase	Hardware Products Reviewed	Baseline Established
Requirements Review	Requirements Definition	Requirements Specification	Requirements
Preliminary Design Review	Design (Preliminary)	Preliminary Design Specification Preliminary Build Plan	—
Critical Design Review	Design (Detailed)	Detailed Design Specification Build Plan Test Plan	Development
Build Design Review	Implementation	Hardware Build	—
Functional Configuration Audit	Testing	CI and its Documentation	Product
Physical Configuration Audit	Testing	CI and its Documentation	

Fig. 3. Hardware reviews.

184

3.2.1 Planning Hardware Reviews

Project and quality management are responsible for initiating formal design reviews. The following personnel are all possible participants in reviews: project manager, project engineer, design engineer, programmers and representatives from quality, technical publications, reliability, production and assembly, and procurement.

3.3 DESIGN REVIEW CHECKLISTS

3.3.1 Requirements Review

This review is held at the end of the requirements definition phase of the hardware life-cycle. At this point the detailed CI requirements specification should be complete and available to those involved.

The following topics should be covered by the review:

—review objectives
—system requirements allocated to the CI
—functional design model
—top-level functional block diagram
—identified external interfaces
—functional requirements
—performance requirements
—physical requirements
—availability, maintainability and reliability
—safety requirements
—security requirements
—compliance matrix
—results of studies and analyses

3.3.2 Preliminary Design Review

This is a formal technical review covering the basic design approach. Schedule the review when the functional design is complete and the preliminary CI design specification is available. At this review examine the design to verify that it meets the requirements in its requirements specification. Ensure that necessary interfaces with other equipment, programs or facilities are identified.

The following topics should be covered at the review:

—a review of the CI requirements
—a summary of deliverables (equipment, spares, documentation)
—an overview of the functional design
—functional block diagrams
—results of studies and analyses
—interface descriptions
—reliability considerations and predictions
—preliminary cabinet elevations
—rough control panel layouts
—preliminary mechanical concepts
—installation design criteria
—test philosophy and requirements
—schedule status

3.3.3 Critical Design Review

This is a formal technical review of the detailed design. It is conducted when the design is functionally complete and detailed block diagrams are complete. Detailed schematic diagrams should be complete and available for review but are not explicitly reviewed at this time.

Detailed fabrication and assembly documentation need not have been prepared by the time of this review; however, the mechanical and packaging concepts should be defined and should be reviewed.

At the time of the review of the CI detailed design specification, the final CI build plan, and the CI test plan, should be complete. The concepts contained in these documents should be explained at the review.

The following topics should be covered at the CI review:

—a review of the CI requirements
—a summary of deliverables (equipment, spares, documentation)
—an overview of the functional design
—functional block diagrams for each hardware assembly
—detailed block diagrams for each board or circuit
—detailed flow diagrams for each firmware program
—detailed interface descriptions
—reliability considerations and predictions
—cabinet and console configurations
—detailed control panel layouts
—detailed mechanical concepts

—installation design criteria
—overview of test plans and resource requirements
—schedule status

3.3.4 Build Design Review

Each build, except the first, should start with a review. Unless changes were made to the CI design requirements or specifications, the review is conducted as an internal review and the format of the presentation may be less formal than the critical review. Should the design have changed in response to changes in system requirements, the build review should be treated in the same manner as a critical review, with the emphasis being on the changes.

The following topics should be addressed at the review:

—requirements to be met in the build
—hardware and firmware elements included in the build
—design modification made during the checking and testing of the previous build
—changes to the scope of the build
—unresolved problems from previous builds to be included in this build
—capabilities of other CIs required by this build (dependencies)
—risks associated with the build
—build test plan
—build schedule

3.3.5 Configuration Audits

At the conclusion of the CI test phase and after the final build test, a functional and physical configuration audit needs to be carried out. These audits are performed by quality management.

Functional configuration audits are performed to verify that the CI meets all of its functional specifications. This review is conducted after the build test and prior to the physical configuration audit. The latter is to ensure that the CI physically agrees with the applicable documents and drawings.

COMPANY B OR C
QUALITY PROCEDURE QPS 007
SUBCONTRACT EQUIPMENT AND SOFTWARE

Compiled by...

Checked by...

Approved by...

Date...

CHANGE HISTORY PAGE

Document status	Date issued	Number of pages	Changed pages	Change/ defect no.
XX	XX	XX	XX	XX

CONTENTS LIST

1 INTRODUCTION

1.1 PURPOSE

To provide guidance on subcontracted work and subcontractors to Companies B or C and the purchase of proprietary software.

1.2 SCOPE

All subcontracted work, proprietary software and developed software. All contractual actions relating to this procedure must be carried out in conjunction with QPS/C 009—Purchasing Quality Procedures.

1.3 TERMS AND ABBREVIATIONS

None.

2 APPLICABLE DOCUMENTS AND REFERENCES

BS 5750 (Part 1) 1987
QPS/C 009—Purchasing and Suppliers
QPA/B/C 011—Goods Inwards and Inspection
QPB 004/QPC 005—Software Design and Standards
QPB 015/QPC 015—Test

3 SELECTION OF SUBCONTRACTORS

The purpose of this section is to provide guidelines so that the capability of potential subcontractors or software suppliers can be assessed prior to placing a contract. Such an assessment may be based on past experience, outside sources of information or direct assessment carried out in the light of the particular requirements. Full details of vendor assessment can be found in QPS/C 009—Purchasing and Suppliers.

Once an order has been placed it is essential that progress is monitored against the requirements and time-scales. Care must be taken to ensure that subcontractors are able to understand and implement the required quality conditions and meet any specific contract requirements. Subcontracts will normally fall into two categories:

—Subcontracted projects that are put out to tender.

—Subcontracted staff brought in to carry out specific tasks (e.g. hardware design, programming).

In both cases the selection will involve the project and engineering and quality manager.

3.1 SUBCONTRACTED PROJECTS

The following items must be considered:

—selection
—specification/contract conditions
—project monitoring
—acceptance
—general points

3.1.1 Selection

The following questions should be addressed:

—Has the proposed subcontractor a sound business and financial base?
—Will the work be carried out at the subcontractor's site or at Company B or C?
—If the former is the case then is there adequate storage for software items together with adequate procurement of hardware and adequate test facilities?
—Who is responsible for quality assurance at the subcontractor?
—Does the subcontractor have his own hardware or software standards and are they suitable for the contract?
—What support/training/warranty will the subcontractor provide?
—What security arrangements exist for the protection of Company B's or C's confidential information?
—Does the subcontractor have the expertise to carry out the contract?

3.1.2 Specification/Contract Conditions

Particular attention should be paid to:

—the adequacy of the requirements specification in allowing sub-contractors to tender;
—the completeness of the subcontractor's response;

191

—the need for a subcontractor quality plan;
—the need for a subcontractor test plan;
—criteria for acceptance from the subcontractor.

3.1.3 Project Monitoring

Design and development reviews should apply to any subcontracted project. The following must be agreed at order placement:

—an agreed programme of subcontractor audits;
—an agreed plan of reviews.

3.1.4 Acceptance

The normal test arrangements should apply to subcontracted work and attention should be given to:

—post-delivery activities, especially change control;
—the date from which warranty is effective;
—approval of documentation (e.g. handbooks);
—the acceptance plan and schedule.

3.1.5 General Points

—Company B or C will maintain records of subcontractors used and this information will be available for selection of subcontractors for new projects.
—It is essential to establish that Company B's or C's customer approves of any subcontractors.
—Purchasing documents must clearly specify the work and any special conditions.

3.2 SUBCONTRACTED STAFF

Selection will normally be carried out by the project or engineering and quality manager or team leader. Candidates will be interviewed as to their technical capability. References will be sought wherever possible. Commercial and purchasing policies for these situations will cover rates, length of contract, termination conditions on both sides and any other specifics. The project or quality manager will be responsible for ensuring that subcontracted staff are aware of any conditions that apply to them.

4 PROPRIETARY/EXTERNAL SOFTWARE

It may be necessary to purchase software from an outside supplier. This may involve complete operating systems or specific packages and may be for one specific project or for more generally used development systems (e.g. compilers). The following should be checked:

—How will the item be accepted and by whom?
—Does the manufacturer/subcontractor provide any test aids to assist in acceptance?
—What training/warranty/maintenance facilities are provided and at what extra cost?
—Is installation included in the price?
—What are the licensing conditions and do these affect Company B's or C's contract with a customer?
—Is there a time limit on the manufacturer's/subcontractor's support?
—What are the delivery time-scales?
—Does purchase of the item imply any specific hardware requirements (e.g. additional memory, co-processors)?

Reports will be kept and reviewed by quality in respect of the above.

COMPANY A OR B

QUALITY PROCEDURE QPS 009

PURCHASING AND SUPPLIERS

Compiled by...

Checked by...

Approved by...

Date...

CHANGE HISTORY PAGE

Document status	Date issued	Number of pages	Changed pages	Change/ defect no.
XX	XX	XX	XX	XX

CONTENTS LIST

1 INTRODUCTION

1.1 PURPOSE

This procedure has been prepared in order to provide written instructions for supplier assessment. Its mandatory requirements shall be implemented by the engineering and quality section, purchasing/administration and, wherever necessary, the managing director.

1.2 SCOPE

This procedure shall apply when assessing all suppliers of services, goods and materials to Company A or B, and to customer-supplied equipment.

The term 'supplier' shall include all sources other than Company A or B that are responsible for providing raw material, part-finished or complete components and services on a commercial basis.

1.3 TERMS AND ABBREVIATIONS

None.

2 APPLICABLE DOCUMENTS AND REFERENCES

QPA/B 011—Goods Inwards and Inspection
QPS 001—Sales and Contract Control
QPA/B 017—Non-conformance, Corrective Action and Records

3 SUPPLIER ASSESSMENT

3.1 ASSESSMENT METHOD

The accepted methods of assessing and approving a potential supplier are:

(1) by evaluating the information returned on questionnaires (see Section 6);
(2) by reputation and previous goods or service;
(3) by accepting the approval of national authorities;
(4) by audit of the supplier's resources.

SUPPLIER QUALITY ASSURANCE RATING Notification to Procurement Department		
	No. of pages	
	Page no.	

Name of Company:

Address of Company:

Contact: Tel.   Dept.

Products/services evaluated:

Basis of evaluation:	
Quality Audit	
Answers to Questionnaire	
Goods Inwards Inspection Results	

Quality Engineer

Quality Rating:

GREEN—Full approval
BLUE—Limited approval
RED—Entirely unsuitable

Remarks:

Signature: ....
(Quality Engineer)

Fig. 1. Supplier quality assurance rating form.

The type and extent of any evaluation shall depend upon the nature of the goods or services to be provided and the degree of previous experience with that supplier.

Suppliers of major products may be subjected to an initial audit to establish their quality assurance capability. They will then be monitored by periodic surveillance.

The results of these evaluations shall be reported on the supplier quality assurance rating form (see Fig. 1), which will be filed in the engineering and quality section.

The form shows, by colour coding, the category of goods or services evaluated, and records the assessment as either fully approved, meeting the requirements of BS 5750 (Parts 1, 2 and 3), or other approval, meeting the above requirements as fully approved but only approved for a specific product or service, unsuitable and non-compliant with the above categories.

A list of approved suppliers is held by the following:

(a) engineering and quality manager
(b) managing director
(c) administration/purchasing

The preferred suppliers shall be subjected to annual review and the approved suppliers' list updated. Companies may be removed from the list, or re-evaluated at any time, at the discretion of the engineering and quality manager, usually when the supplier's performance remains in a Class C category over the annual period. The performance categories are:

Class A—excellent; non-conformance is rare.
Class B—fair; non-conformance is noticeable or has worsened.
Class C—bad; non-conformance causes significant problems.

Secondary source suppliers are assessed and categorised as above according to their performance but on receipt of goods.

4 QUALITY ASSESSMENT

Where a supplier has been officially audited by Company A or B then a complete quality assurance questionnaire (see Section 6), along with a report concerning the audit, shall be prepared by the engineering and quality manager and a copy forwarded to the supplier.

The engineering and quality manager will receive details of all rejections, site reports and complaints where they concern the quality of material or service supplied to Company A or B.

4.1 INFORMATION GATHERING

The person identifying a non-conformance shall be responsible for supplying the managing director and engineering and quality manager with details, who will update their records. Such non-conformances shall include, for example:

(a) scrap and replace
(b) scrap and not replace
(c) rectify at supplier's works
(d) rectify at Company A or B

The appropriate non-conformance shall be identified with the purchase requisition number and production documentation. This will contain the supplier's details relevant to the material rejected.

Certain materials are subject to inspection on receipt. For such materials goods inwards shall follow the goods inwards procedures and forward all details to administration/purchasing and quality.

4.2 RECORDS

The engineering and quality manager shall keep a record, for 3 years, of all associated documentation concerning suppliers. These shall include production data, quality questionnaires, audit reports and any other relevant information.

5 EXAMPLE OF SUPPLIER ASSESSMENT CHECKLIST FOR SUBCONTRACTORS AND STOCKIST ORGANISATIONS

5.1 SUPPLIER ASSESSMENT CHECKLIST FOR SUBCONTRACTORS

(1) The results of the supplier evaluation shall be reported on Fig. 1, which will be filed in the engineering and quality section.
(2) Does the contractor have a quality system that meets the requirements of BS 5750 (Parts 1, 2 and 3)?
(3) (a) Does the contractor have an individual responsible for quality/ inspection?

199

   (b) Is he independent of other functions (i.e. manufacturing and production)?

   (c) Does he have the authority to enable him to resolve all quality inspection matters?

(4) (a) Does the contractor have adequately documented procedures for inspection and quality (i.e. quality manual, written work instructions)?

   (b) If not, how does he ensure vital inspection and quality functions are carried out?

   (c) Does the contractor have a system for the formal updating and revision of standards and work instructions vital to the quality of the products?

   (d) Does the contractor maintain inspection and test records for a prescribed period?

   (e) Do the records verify that essential test/performance have been carried out?

   (f) Does the contractor have a recognised corrective action procedure?

(5) Does the purchasing documentation contain clear descriptions of the products required and any necessary supporting documentation?

(6) Does the contractor have procedures which ensure that only the latest applicable technical data and/or drawings are used?

(7) (a) Does the contractor's incoming inspection ensure that only acceptable items are used for manufacture or distribution?

   (b) Are non-conforming items at the goods inwards, assembly test and inspection stages identified with a rejection tag and segregated?

   (c) Are records of non-conforming items and their disposition maintained?

(8) Does the contractor have a procedure for the return of non-conforming goods (i.e. rejection notes, rejection tag)?

(9) (a) Are all inspection and test devices in a known state of calibration?

   (b) Is documentary evidence available?

   (c) By what method does the contractor determine the frequency of calibration of inspection and test devices?

(10) (a) Does the contractor keep inspection and test records?

   (b) If so, for how long?

(11) Does the contractor have a review and evaluation system to ensure all quality functions are kept up to date?

(12) Is there any final inspection of goods carried out before despatch to ensure conformity with order?

(13) Are there any checks made on special requirements for preservation and packaging before despatch?

5.2 SUPPLIER ASSESSMENT CHECKLIST FOR STOCKIST ORGANISATIONS

(1) The results of the supplier evaluation shall be reported on Fig. 1 (supplier quality assurance rating), which will be filed by the engineering and quality manager.

(2) (a) Does the stockist have an individual responsible for quality/inspection?

    (b) Is he independent of other functions?

    (c) Does he have the authority to enable him to resolve all quality inspection matters?

(3) (a) Does the stockist have adequately documented procedures for inspection and quality (i.e. quality manual, written work instructions)?

    (b) If not, how does he ensure vital inspection and quality functions are carried out?

    (c) Does the supplier have a system for the formal updating and revision of the standards and work instructions vital to the quality of the product?

(4) Does the purchasing documentation contain clear descriptions of the products required and any necessary supporting documentation?

(5) (a) What is the stockist's system of segregation and identification of purchased products?

    (b) Do any goods have a limited shelf-life and need any special protection?

(6) What checks are made on incoming goods to ensure that only acceptable items are placed in stock?

(7) Are all non-conforming items clearly identified with a reject tag and segregated?

(8) What procedure is adapted for the return of non-conforming purchased goods, and is this procedure documented (i.e. rejection notes or rejection tags)?

(9) (a) Does the stockist have his own testing facilities?

    (b) Alternatively, can the stockist supply test certificates?

201

(10) (a) Are all inspection and test devices in a known state of calibration?

(b) Is documentary evidence available?

(c) By what method does the stockist determine the frequency of calibration of inspection and test devices?

(11) (a) Does the stockist keep inspection and test records?

(b) If so, for how long?

(12) Does the stockist have a review and evaluation system to ensure all quality functions are kept up to date?

(13) Is there any final inspection of goods carried out before despatch to ensure conformity with order?

(14) Are any checks made on special requirements for preservation and packaging before despatch?

6 EXAMPLE OF QUALITY ASSURANCE SUPPLIER EVALUATION QUESTIONNAIRE

The following information is required for our supplier evaluation system. Please complete the questionnaire by ticking the appropriate column.

1. Name and address of company..

....

....

2. Tel. No... Fax No...

3. Give brief details of products/services which your company may supply to Company A or B.

....

....

....

4. Has your company's quality organisation been approved by any national quality/inspection authority such as CEGB, MOD, BNFL or NNC? If yes, give your registration number and approval standard.

5. Do you have an established quality control/inspection department within your company?

6. Is there a senior person responsible for the quality of the goods leaving your works?

7. Does your company have a system for ensuring that the products supplied are of an acceptable quality and also meet the purchaser's contractual requirements?

8. Which of the following inspection/testing services can you provide? Examples include:

    (a) goods inwards
    (b) first-off or sample inspection
    (c) final inspection
    (d) inspection at despatch
    (e) NDT services
    (f) others (please state)

9. Can you supply one of the following certificates which might be required for certain items on our contracts?

    (a) material certificates
    (b) letter of conformity or guarantee
    (c) functional test certificate

10. Do you have a formal procedure for ensuring that only up-to-date drawings/standards are used in your company?

11. Are all incoming orders checked on receipt to ensure that the drawings/standards quoted are the ones in your possession?

12. Do you have an established procedure for the regular calibration of your measuring and test equipment, which is essential to the maintenance of quality?

13. Does your company keep records of all important tests, etc., for at least five years?

14. Give the name of the person to be contacted in connection with the questionnaire?

Signed ....
Company A or B

7 EXAMPLE OF LETTER TO BE SENT TO SUBCONTRACTORS

A letter similar to the one below may be compiled and sent to the subcontractors by the engineering and quality manager.

Dear Sirs,

SUBCONTRACTORS' ASSESSMENT

A number of our major customers are introducing into their contracts requirements concerning quality assurance. It is their ultimate intention to use only firms with satisfactory quality assurance, and to compile a list of approved suppliers.

It is necessary to control the quality of all purchased goods and services provided for all of our contracts. To comply with this particular requirement it has been necessary to establish our own list of approved suppliers and develop a system to verify, by questionnaire, visits and records of performance, the abilities of our suppliers.

In order, therefore, that records can be established or updated for your company, I would be grateful if you would complete the attached questionnaire as soon as possible and return it to the address printed at the top of this letter.

Yours faithfully,

.... Quality Manager

8 PURCHASING

The engineering and quality manager or managing director will check the vendor rating records maintained by quality. The managing director will liaise with quality before the purchase requisition is allowed to proceed.

Quality and stores will allocate material required from any available free stock, and identify the exact quantities that need to be purchased, having checked out if the item is already on order.

The purchase requisition and quotation is then passed to the managing director for approval. Administration enters the details of the purchase requisition into the computerised order system. Each purchase requisition is uniquely identified. The purchase requisition is generated, checked and signed before being posted to the supplier. In many instances the supplier's detailed catalogue number uniquely specifies the purchased item and its standards.

Purchase requisition copies are distributed, as necessary, to:

(1) goods inwards
(2) purchasing/administration
(3) engineering and quality

Administration/purchasing monitors the status of all purchase orders. Materials received at goods inwards are dealt with under procedure QPA/B 011.

9 CLIENT SUPPLIED/LOANED EQUIPMENT

Where client supplied/loaned equipment is used on a project, the engineering and quality manager will be responsible for keeping a record of all materials (normally computer hardware or test equipment) provided on loan for use on Company A's or B's premises. In addition, the engineering and quality manager will be responsible for ensuring that the following aspects are agreed with the customer and that the agreed conditions are included in the quality plan or other project document. These include:

—responsibility for providing and paying for maintenance to the equipment;
—responsibility for providing and paying for insurance;
—provision of and payment for consumables (media, paper, etc.);
—Company A's or B's acceptance of equipment;
—acceptance criteria;
—acceptance signatures.

COMPANY A OR B

QUALITY PROCEDURE QPS 012

IDENTIFICATION AND TRACEABILITY

Compiled by...

Checked by...

Approved by...

Date...

Note: The detail of this procedure relates more closely to the QPAs rather than the QPBs. This should be taken into consideration when designing your procedures.

CHANGE HISTORY PAGE

Document status	Date issued	Number of pages	Changed pages	Change/ defect no.
XX	XX	XX	XX	XX

CONTENTS LIST

1 INTRODUCTION

1.1 PURPOSE

To provide a procedure for identifying the status of all items at all stages.

1.2 SCOPE

All items within Company A or B.

1.3 TERMS AND ABBREVIATIONS

C of C—certificate of conformity

2 APPLICABLE DOCUMENTS AND REFERENCES

None.

3 PROCEDURE

All cartons and boxes will be individually labelled with a tag. All large items will be individually labelled with a tag.

The tags will be physically fastened to the item when received at goods

---

Description..

....

Supplier...

Our Order No...

C of C Number...

Date of Arrival ....

Fig. 1. Example of tag.

inwards. The tag will remain on the item throughout its movement in the factory.

An example of the tag is shown in Fig. 1. C of C refers to certificate of conformity as supplied by the supplier.

4 ROUTING DOCUMENTATION

The method of identifying items and their quality and test status is by use of a routing production folder.

When each item is received at Company A or B a routing folder will be raised by administration. It will consist of each of the forms shown in QPA or B 010.

The folder consists of an envelope or file; it will accompany the item at all times until customer acceptance. To achieve this the equipment serial number will be recorded on both the equipment itself and the routing documents.

After customer acceptance the folder will be filed, by contract or order number, in the engineering and quality department. This procedure will also apply to purchaser-supplied equipment.

COMPANY A OR B
QUALITY PROCEDURE QPS 013
STORES

Compiled by..

Checked by..

Approved by..

Date..

Note: The detail of this procedure relates more closely to QPA 010 than to QPB 010.
This should be taken into account when designing your procedures.

CHANGE HISTORY PAGE

Document status	Date issued	Number of pages	Changed pages	Change/ defect no.
XX	XX	XX	XX	XX

CONTENTS LIST

1 INTRODUCTION

1.1 PURPOSE

To provide a procedure for the control of items in and out of stores.

1.2 SCOPE

Company A or Company B stores.

1.3 TERMS AND ABBREVIATIONS

GIN—goods inwards note
ISO—internal sales order
C of C—certificate of conformity
MR—material requisition

2 APPLICABLE DOCUMENTS AND REFERENCES

QPA/B 010—Control of Production

3 HOUSEKEEPING

At all times the stores must be maintained in a tidy and clean condition, and all piece parts and components properly identified.

4 RECEIPT OF GOODS INTO STORES

The receiving storeman will check that the goods received are consistent in respect of quantity and description as advised on the accompanying GIN. Having satisfied himself that all is in order, he will then locate the goods on the appropriate shelves and file a copy of the GIN for records. No goods will be received into stores unless the GIN has been signed by the inspector (see Fig. 1 for GIN).

When goods or materials are received under cover of the supplier's release note or C of C, details should be noted in the C of C register showing the name of the supplier, the quantity received and the date or batch code.

Order No. ....

| METHOD OF DELIVERY | COMPLETED ORDER | PART ORDER | BONDED STOCK | YES/NO | NARRATIVE |
| ADVICE NOTE No. | RECEIVED BY | | SUPPLIER | | |

| ITEM No. | ADVISED QUANTITY ACCEPTED | UNITS | CONSIGN No. | UNIT PRICE | DESCRIPTION / PART NUMBER | SHORTAGE | REJECT NOTE N |
| 9 10 11 12 13 14 15 16 17 18 19 20 21 22 23 24 | 25 26 27 28 29 30 31 32 33 | 45 46 47 48 49 50 51 52 53 54 55 56 57 58 59 60 61 62 63 64 65 66 67 68 69 70 71 | | | | | |

PASSED TO STORES BY

Q A ACCEPTED BY

STORES CHECKED AND ACCEPTED

DEPT.

SIGNED

Fig. 1. Goods inwards note.

QPS 013
Issue 1

213

Requisition no. .... Date ....

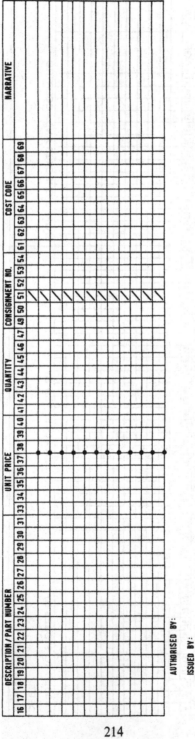

Fig. 2. Material requisition.

5 ISSUE OF GOODS FROM STORES

The storeman will, on receipt of a request for material, etc., locate the goods and issue them to the appropriate department on a material requisition form (Fig. 2), which will be signed by the recipient.

In the case of approved components the storeman will consult the C of C register and issue components bearing the earliest date code.

Details of withdrawal and ISO or material requisition number will be entered in the register. Any component bearing date codes older than three years must not be used on some contracts. Where special commercial use only is indicated, or if stockist's approval has been obtained, they shall be returned to the supplier for retest if the three years is exceeded.

6 STOCK ROTATION

It is good stores practice to ensure parts are controlled so as to ensure that they are issued on a 'first in, first out' basis. In order to assist in this rotation relevant piece parts' containers will have the goods inwards note number recorded on them. This also gives a cross-reference back to the particular supplier.

6.1 STOCK ISSUES

Issue can only be effected on a material requisition form. All paperwork leaving the stores must agree both in quantity and description with the goods issued. When the stock level of bulk issue items falls the storeman will, by use of a minimum stock level, raise a purchase requisition as outlined in QPS 009.

7 GOODS INWARDS NOTE

This is used to document the movement of items from the goods inwards to stores.

It is important that the following points are strictly observed.

- All writing must be in ball-point pen.
- Any errors should be struck through with a single line and signed by

the person making the adjustment. The revised figure should be clearly written in.

- Any spoilt GINs should be 'cancelled' written across them and a copy routed to administration.
- The section 'description' should have the correct code written in. This is most important.
- The section 'quantity' should have the quantity advised written in.
- The section 'quantity accepted' should be filled in only after the engineering and quality department have tested items.
- The sections 'method of delivery' and 'advice note no.' should be completed by goods inwards, who should sign and date the document in the spaces provided.
- The storeman should sign all copies of the GINs on receipt of the correct items.
- The section 'part number' should have the relevant information entered therein.

8 MATERIAL REQUISITION (Fig. 2)

This is used to record the movement of items from stores to production. It is important that the following points are strictly observed.

- All writing should be in ball-point pen.
- Any errors should be struck through with a single line and signed by the person making the adjustment, and the revised figure clearly written in.
- Any spoilt MR should have 'cancelled' written across it and a copy routed to administration. Also a copy should be filed in stores.
- The sections 'descriptions' and 'quantity/weight' shall be completed by the storeman issuing the goods.
- The section 'authorised by' shall be signed by the storeman.

9 PURCHASE ORDER CHASING

Purchasing/administration will review outstanding purchase requisitions on a weekly basis. Goods not received by the quoted delivery date will be progressed.

216

10 KITTING

10.1

On receipt from administration of the production folder (see QPA/B 010) containing documents necessary for production of items, stores will proceed to withdraw the parts from stock as indicated on the enclosed parts list (see Fig. 3).

Parts list as appropriate to your company

Fig. 3. Parts list.

10.2

If all parts are available as per the parts list then stores will locate and allocate any discrete part as indicated on the design sheet, then passing the completed kit of parts, together with the production folder, to the production department. Stores will retain one copy of the parts list for record purposes.

10.3

Should there be any parts or discrete products not available, a purchase requisition should be requested via purchasing/administration or a works order number (see Fig. 1) raised on production.

11 SECURITY

The stores area shall be physically segregated and access restricted to authorised persons.

COMPANY A OR B
QUALITY PROCEDURE QPS 016
CALIBRATION OF MEASURING EQUIPMENT

Compiled by...

Checked by...

Approved by...

Date...

CHANGE HISTORY PAGE

Document status	*Date issued*	*Number of pages*	*Changed pages*	*Change/ defect no.*
XX	XX	XX	XX	XX

CONTENTS LIST

1 INTRODUCTION

1.1 PURPOSE

The purpose of this document is to describe the procedure for ensuring and maintaining an effective system for the control and calibration of equipment in accordance with BS 5750.

1.2 SCOPE

The scope of this document applies to the control and calibration of measurement standards and measuring equipment being used within Company A or B.

Calibration shall be carried out only by trained personnel. Only equipment/tools purchased by Company A or B shall be used.

1.3 TERMS AND ABBREVIATIONS

None.

2 APPLICABLE DOCUMENTS AND REFERENCES

Manufacturer's specifications for equipment specified in Section 3.

3 EQUIPMENT

3.1 CALIBRATED EXTERNALLY

The following equipment is used for in-house calibration:

(a) calibrated multimeter—RS Fluke 77
(b) frequency source
(c) temperature probe
(d) EPROM programmer—Stag PP39

3.2 CALIBRATED IN-HOUSE

The following equipment is calibrated using the above equipment:

(a) multimeter—Fluke 3·5 digit DMM

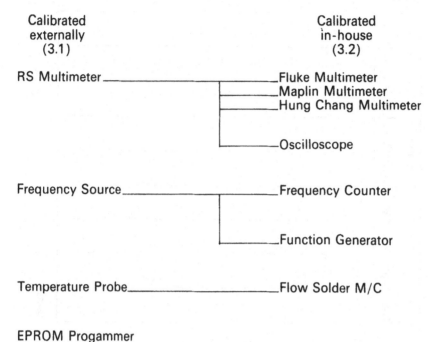

Calibrated
externally
(3.1)

Calibrated
in-house
(3.2)

RS Multimeter ———————————— Fluke Multimeter
—————— Maplin Multimeter
—————— Hung Chang Multimeter

—————— Oscilloscope

Frequency Source ———————— Frequency Counter

—————— Function Generator

Temperature Probe ————————— Flow Solder M/C

EPROM Progammer

Fig. 1. Traceability of measurements.

    (b)   multimeter—Maplin 3·5 digit DMM
    (c)   multimeter—Hung Chang 4·5 DMM
    (d)   frequency counter—Black Star Meteor 100
    (e)   function generator—Black Star Jupiter 500
    (f)   oscilloscope—Hameg 201
    (g)   flow solder machine including solder bath and pre-heat ther-
          mometers/thermostats
    (h)   power supplies

Figure 1 shows the measurement traceability whereby in-house calibrated items are checked against the externally calibrated items.

4 CALIBRATION AND EQUIPMENT MASTER REGISTER

A master list of all equipment is kept in the 'calibration and equipment master register' (see Fig. 2). It details the description of the equipment, serial

Description	Serial No.	Period	1991				1992								
			Sept.	Oct.	Nov.	Dec.	Jan.	Feb.	Mar.	Apr.	May	June	July	Aug.

Fig. 2. Calibration and equipment master list.

Instrument..

Serial No...

Manufacturer...

Approved signature

Company A or B

Date

Fig. 3. In-house calibration certificate.

numbers and the calibration period. All new equipment shall be recorded in the master calibration and equipment register; also all calibration certificates shall be kept in the register (see Fig. 3).

The calibration and equipment register is maintained and kept in the engineering and quality section.

The calibration and equipment register shows when equipment becomes due for calibration, detailing the description of the equipment, serial numbers and the due calibration month.

When the equipment becomes due for calibration, the equipment user is notified in advance and the equipment is made available for calibration.

It may be necessary to extend the calibration period if the equipment is required urgently. Refer to Section 8, Variation of Calibration Period.

5 INTERVAL OF CALIBRATION

Measurement standards and measuring equipment shall be calibrated externally at an interval of one year unless specified otherwise.

A review of the calibration interval for a particular piece of equipment will be carried out by the engineering and quality manager, and the interval increased or decreased as equipment trends dictate.

In-house calibrated items will initially be calibrated at six-monthly intervals.

6 APPROVED TEST HOUSE (CALIBRATION)

If the amount of equipment requiring calibration increases then a test house approved to Defence Standard 05-26 will be used.

7 LABELLING, SEALING AND RECORDING

7.1 LABELLING

All equipment shall carry a serial number label and calibration label showing the date when the next calibration is due (see Fig. 4).

CALIBRATED

By _____ Date _____

Due _____

Fig. 4. Calibration label/seal.

7.2 SEALING

Access to adjustable devices on measurement standards and measuring equipment, which are fixed at the time of calibration, shall be sealed or otherwise safeguarded to prevent tampering by unauthorised personnel. Seals shall be designed such that tampering shall destroy them.

7.3 RECORDING

The calibration results for all equipment shall be recorded on the 'calibration master register' (see Fig. 2) or, in the case of an outside test-house, on a calibration certificate, which must be dated and signed with fully detailed remarks or comments.

8 VARIATION OF CALIBRATION PERIOD

It may become necessary to retain equipment in an emergency, which may fall outside the calibration period. In which case the equipment may have its calibration period extended by a further month, but only if the previous calibration records show that there was no major deterioration in the accuracy of the equipment.

If the equipment is satisfactory it shall be labelled up showing the date of the extended period. This shall also be noted in the records.

The calibration period (see Section 5) may be lengthened or shortened by the engineering and quality manager, providing the trend in the previous three measurements justifies a change. Details of the change of periodicity will be entered on the master register.

9 INVALIDATION OF CALIBRATION

All equipment shall be immediately removed and calibrated if the equipment

(a) is outside the designated calibration period (see Section 8);
(b) failed in operation;
(c) goes outside limits;
(d) is damaged in a way that may affect accuracy.

10 STORAGE AND HANDLING

All equipment shall be protected against any damage, misuse or change in dimensional or functional characteristics during transporting and storing.

11 HIRED EQUIPMENT

In an emergency it may be necessary to hire equipment. In this instance it is vital that the hired equipment shall be in a known state of calibration and labelled up accordingly, and where possible a calibration certificate is supplied by the hire company.

COMPANY A OR B
QUALITY PROCEDURE QPS 018
QUALITY RECORDS

Compiled by..

Checked by..

Approved by..

Date..

CHANGE HISTORY PAGE

Document status	Date issued	Number of pages	Changed pages	Change/ defect no.
XX	XX	XX	XX	XX

CONTENTS LIST

1 INTRODUCTION

1.1 PURPOSE

To define the responsibilities for the retention and maintenance of quality records in accordance with the requirements of Companies A or B and their customers.

1.2 SCOPE

This procedure details the records that will be kept by the engineering and quality department.

It applies to all contracts and orders placed on Company A or B. It sets out requirements for records that must be kept in connection with all items of hardware and software.

1.3 TERMS AND ABBREVIATIONS

None.

2 APPLICABLE DOCUMENTS AND REFERENCES

QPS 001—Contract Review
QPA/B 010—Control of Production
QPS 009—Purchasing and Suppliers
QPS 016—Calibration of Measuring Equipment
QPA/B 011—Goods Inwards and Inspection
QPA/B 019—Internal Quality Audit (also Subcontractor Audit for QPB)

3 PROCEDURE

All the records that are defined in Section 4 shall be maintained for all systems hardware and software. These records will be maintained by the engineering and quality department, in numerical, date or contract number order as applicable, for a period of five years. It is the engineering and quality manager's responsibility to ensure that records are complete and accurate in order that they can form a basis for analysis and management action.

4 RECORDS MAINTAINED

The records shall consist of:

4.1 CONTRACT

A copy of all contracts/orders and associated amendments shall be made available to the engineering and quality manager by the sales manager. The engineering and quality department will ensure that these are incorporated into the records.

4.2 PRODUCTION FOLDER

The completed production routing and inspection details shall be retained by the engineering and quality department in numerical order.

4.3 CONCESSION/DEFECT, CHANGE REQUEST, ERROR/CHANGE REPORT

These are the primary methods of reporting significant technical difficulties with hardware or software before, during or after commissioning. They permit work to continue pending investigation. The problem may involve:

—design problems
—defective equipment
—drawing errors or inaccuracy
—incorrect item selected or delivered
—incorrect documentation
—deviation from drawing or specification
—use of non-approved spares
—interface incompatibility

All forms and reports must be numbered consecutively and cross-referenced to the appropriate contract. The engineering and quality manager will discuss, with the appropriate manager, each problem and will agree the course of remedial action. Engineering and quality department will follow up all remedial action to ensure that it is effective.

4.4 PURCHASER AND SUPPLIER

These records include those described in QPS 009 (Purchasing and Suppliers). Where visits to suppliers result in reports, these will be included in the records.

4.5 AUDIT REPORTS

These fall into three categories:

—internal quality management reports
—internal project quality audit reports
—subcontract audit reports

From these, problems will be dealt with using the methods listed in Section 4.3.

4.6 CERTIFICATES OF CONFORMITY

Wherever they have been requested, certificates of conformity will be retained for all procured items.

4.7 CERTIFICATES OF DESIGN

Certificates of design will be obtained, when required, from subcontractors.

4.8 QUALITY DOCUMENTATION

Amendments to all quality documentation will be distributed to all holders of the document in question as per the distribution list. Suppliers' quality documentation will be retained.

4.9 CALIBRATION RECORDS

All calibration sheets and measurements will be retained. QPS 016 (Calibration) provides details.

4.10 PROJECT ARCHIVES

At the completion of each contract all the project records (e.g. specifications, drawings, inspection results) will be retained. They will be filed by project name or number.

4.11 GOODS INWARDS

All goods received documentation raised against goods will be maintained by the goods inwards inspection department and later archived.

4.12 MINUTES OF MEETINGS AND REVIEWS

Minutes of meetings and reviews will be filed under their respective headings and in date order.

4.13 OTHER RECORDS

Other records will be kept as appropriate under contract conditions.

COMPANY A, B OR C
QUALITY PROCEDURE QPS 020
EDUCATION AND TRAINING

Compiled by...

Checked by...

Approved by ....

Date..

CHANGE HISTORY PAGE

Document status	*Date issued*	*Number of pages*	*Changed pages*	*Change/ defect no.*
XX	XX	XX	XX	XX

CONTENTS LIST

1 INTRODUCTION

1.1 PURPOSE

Training of personnel is an integral part of any business since all staff need to be motivated as well as competent to perform the tasks for which they are employed. Their own career preferences and personal objectives are a part of this.

1.2 SCOPE

Training should be made available to personnel at all levels in all departments. Senior management and technical personnel should be actively encouraged to train and retrain as the nature of the business and products evolves.

1.3 TERMS AND ABBREVIATIONS

None.

2 APPLICABLE DOCUMENTS AND REFERENCES

BS 5750 Part 0 Section 0.2 (ISO 9004: 1987)
BS 5750 Part 1 (ISO 9001: 1987)

3 MANAGEMENT RESPONSIBILITIES

Senior management has a responsibility to ensure that personnel have career plans, adequate education and training for the job, and regular performance reviews to monitor progress against agreed objectives.

There must be an 'open door' policy at all levels, and personnel must be encouraged to approach management with ideas of interest and topics of concern. To this end regular meetings, both formal and informal, are to be encouraged.

3.1 MANAGEMENT AND PERSONNEL

All managers have the following basic responsibilities towards their personnel:

—introducing new personnel and helping them to understand the overall goals and policies of the company;

—communicating the company's policy and procedure in respect of career planning, monitoring and performance appraisal;

—evaluating personnel development and performance;

—informing personnel of relevant courses, seminars, educational opportunities, etc.;

—coordinating an effective needs-based educational programme and monitoring its progress;

—making personnel aware of new project opportunities where these are related to their planned development;

—maintaining a skills database and training records for all personnel;

—ensuring planned performance reviews and implementing appropriate salary action;

—preparing action plans where personnel weaknesses have been revealed by the reviews.

3.2 PERSONNEL AND MANAGEMENT

Personnel have responsibilities towards their managers as follows:

—discussing any problems as they arise;

—keeping management informed about what brings job satisfaction;

—developing themselves along their chosen career path;

—attending courses, etc., when the opportunities are provided;

—identifying short- and long-term career objectives;

—carrying out a self-analysis before each review date so as to maximise their contribution to the review and any action plan.

3.3 MOTIVATION

Motivation is critical to the success of any business. Management must actively encourage staff to take part in quality awareness programmes and to participate with them in improving quality. Managers must be seen to be vigorously involved and committed to such activities.

Where groups of individuals achieve significant quality improvements they must be seen to be recognised and rewarded.

Training and education progression programmes should always involve the needs of the individual, as well as those of the company, to ensure the maximum integration between the two. This can only come from an open communications policy and an active interest by management.

4 TRAINING RECORDS

It is important that training records are kept and that they are regularly updated and reviewed. They have an important role in allowing personnel with the appropriate training and qualifications to be matched against particular roles.

A training file will be opened for each new starter.

5 RECRUITMENT POLICY

A recruitment policy will be established which ensures that competent and well-motivated staff are selected who map to the company's business needs.

COMPANY A OR B
QUALITY PROCEDURE QPS 021
SERVICING

Compiled by...

Checked by...

Approved by...

Date..

CHANGE HISTORY PAGE

Document status	Date issued	Number of pages	Changed pages	Change/ defect no.
XX	XX	XX	XX	XX

CONTENTS LIST

1 INTRODUCTION

1.1 PURPOSE

The purpose of this procedure is to outline the method for dealing with any items returned from site and for their routing and documentation.

1.2 SCOPE

All items returned from site for any reason.

1.3 TERMS AND ABBREVIATIONS

None.

2 APPLICABLE DOCUMENTS AND REFERENCES

QPS 001—Contract Review
QPA/B 011—Goods Inwards and Inspection
QPA 014—Line Inspection and Test
QPB 014—In-line Inspection
QPB 015—Test
QPS 022—Statistical Techniques

3 PRE-INSPECTION

Items returned from site will arrive at goods inwards and be examined by the engineering and quality manager to ensure that they were adequately packed and that damage was not sustained due to transit.

4 BOOKING-IN PROCEDURE

4.1 Concession/Defect Sheet

After pre-inspection the engineering and quality manager will raise and complete the appropriate sections of the concession/defect sheet (Fig. 1).

| Date | Project | Project No. | Sheet No. |

Subject/Item

Raised by: Date:

Details of defect:

Copy to Engineering and Quality Manager

Corrective actions:

Analysis of defect

Details of Warranty/Service Agreement

Defect correctable
Not correctable, concession needed (delete as applicable).

Signed .... (Engineering and Quality Manager)

Date ....

Fig. 1. Concession/defect sheet.

Any documentation from the customer will be attached to the sheet, as will copies of relevant internal documents (e.g. drawings, contract).

4.2 RETURNED ITEMS

All returned items and their associated documentation will be clearly identified and segregated in a non-conforming goods area of goods inwards.

240

5 INVESTIGATIONS

All investigations of field defects and returned goods are the responsibility of the engineering and quality manager, who will ensure that for each item:

—it will be tested for the alleged fault;
—the design or manufacturing process is not at fault;
—if the fault is proved, appropriate remedial action is taken;
—that any resulting changes to process, design or documentation are put in hand;
—the customer is informed of the situation;
—the conditions of the contract in question and any associated warranty are met.

5.1 SERVICE/WARRANTY AGREEMENT REVIEWS

All such agreements will be reviewed annually by the engineering and quality manager.

6 DEFECT RECORDING SYSTEM

All warranty/service records will be maintained and will contain at least the following:

—repair and response times;
—results of the defect and, where relevant, down-times;
—identity of the supplier if the defect arose from a purchased item;
—customer details, including the nature of use of the product;
—the cause of the defect.

Statistics and trends will be kept (QPS 022), and the engineering and quality manager will be responsible for initiating appropriate remedial action whenever a trend is revealed.

7 SPARES AND PERSONNEL AVAILABILITY

Adequate stocks of spare parts must be kept in order to satisfy all warranties and agreements in force.
Personnel must be available and trained to carry out repairs as necessary.

COMPANY A OR B
QUALITY PROCEDURE QPS 022
STATISTICAL TECHNIQUES

Compiled by...

Checked by...

Approved by ....

Date...

CHANGE HISTORY PAGE

Document status	Date issued	Number of pages	Changed pages	Change/ defect no.
XX	XX	XX	XX	XX

CONTENTS LIST

1 INTRODUCTION

1.1 PURPOSE

To provide guidance on sampling techniques for the purpose of inspection.

1.2 SCOPE

This document describes the sampling methods to be used by goods inwards inspection for the inspection of items packed in a number of boxes.

1.3 TERMS AND ABBREVIATIONS

Box—any container, package, etc., containing items to be inspected.

2 APPLICABLE DOCUMENTS AND REFERENCES

QPA/B 011—Goods Inwards and Inspection
BS 6001—Inspection Procedures and Tables for Sampling by Attribute

3 METHOD

Determine number of boxes in batch, and use General Inspection Level II—Normal Sampling to determine number of boxes to be taken as sample (see Fig. 1).

Select sample at random and withdraw boxes from batch for further inspection.

Open each box in sample and verify that contents are correctly identified by the box label.

Unless otherwise specified, inspection will be to 1% AQL (see Fig. 2).

Lot or batch size	Special inspection levels				General inspection levels		
	S-1	S-2	S-3	S-4	I	II	III
2 to 8	A	A	A	A	A	A	B
9 to 15	A	A	A	A	A	B	C
16 to 25	A	A	B	B	B	C	D
26 to 50	A	B	B	C	C	D	E
51 to 90	B	B	C	C	C	E	F
91 to 150	B	B	C	D	D	F	G
151 to 280	B	C	D	E	E	G	H
281 to 500	B	C	D	E	F	H	J
501 to 1 200	C	C	E	F	G	J	K
1 201 to 3 200	C	D	E	G	H	K	L
3 201 to 10 000	C	D	F	G	J	L	M
10 001 to 35 000	C	D	F	H	K	M	N
35 001 to 150 000	D	E	G	J	L	N	P
150 001 to 500 000	D	E	G	J	M	P	Q
500 001 and over	D	E	H	K	N	Q	R

code letters

Fig. 1. BS 6001—Sample size code letters.

Acceptable Quality Levels (normal inspection)

(Values given as Ac Re. ↓ = use first sampling plan below arrow. ↑ = use first sampling plan above arrow.)

Sample size code letter	Sample size	0·010	0·015	0·025	0·040	0·065	0·10	0·15	0·25	0·40	0·65	1·0	1·5	2·5	4·0	6·5	10	15	25	40	65	100	150	250	400	650	1000
A	2	↓	↓	↓	↓	↓	↓	↓	↓	↓	↓	↓	↓	↓	↓	↓	↓	0 1	1 2	2 3	3 4	5 6	7 8	10 11	14 15	21 22	30 31
B	3	↓	↓	↓	↓	↓	↓	↓	↓	↓	↓	↓	↓	↓	↓	↓	0 1	1 2	2 3	3 4	5 6	7 8	10 11	14 15	21 22	30 31	44 45
C	5	↓	↓	↓	↓	↓	↓	↓	↓	↓	↓	↓	↓	↓	↓	0 1	1 2	2 3	3 4	5 6	7 8	10 11	14 15	21 22	30 31	44 45	↑
D	8	↓	↓	↓	↓	↓	↓	↓	↓	↓	↓	↓	↓	↓	0 1	1 2	2 3	3 4	5 6	7 8	10 11	14 15	21 22	30 31	44 45	↑	↑
E	13	↓	↓	↓	↓	↓	↓	↓	↓	↓	↓	↓	↓	0 1	1 2	2 3	3 4	5 6	7 8	10 11	14 15	21 22	30 31	44 45	↑	↑	↑
F	20	↓	↓	↓	↓	↓	↓	↓	↓	↓	↓	↓	0 1	1 2	2 3	3 4	5 6	7 8	10 11	14 15	21 22	30 31	44 45	↑	↑	↑	↑
G	32	↓	↓	↓	↓	↓	↓	↓	↓	↓	↓	0 1	1 2	2 3	3 4	5 6	7 8	10 11	14 15	21 22	30 31	44 45	↑	↑	↑	↑	↑
H	50	↓	↓	↓	↓	↓	↓	↓	↓	↓	0 1	1 2	2 3	3 4	5 6	7 8	10 11	14 15	21 22	30 31	44 45	↑	↑	↑	↑	↑	↑
J	80	↓	↓	↓	↓	↓	↓	↓	↓	0 1	1 2	2 3	3 4	5 6	7 8	10 11	14 15	21 22	30 31	44 45	↑	↑	↑	↑	↑	↑	↑
K	125	↓	↓	↓	↓	↓	↓	↓	0 1	1 2	2 3	3 4	5 6	7 8	10 11	14 15	21 22	30 31	44 45	↑	↑	↑	↑	↑	↑	↑	↑
L	200	↓	↓	↓	↓	↓	↓	0 1	1 2	2 3	3 4	5 6	7 8	10 11	14 15	21 22	30 31	44 45	↑	↑	↑	↑	↑	↑	↑	↑	↑
M	315	↓	↓	↓	↓	↓	0 1	1 2	2 3	3 4	5 6	7 8	10 11	14 15	21 22	30 31	44 45	↑	↑	↑	↑	↑	↑	↑	↑	↑	↑
N	500	↓	↓	↓	↓	0 1	1 2	2 3	3 4	5 6	7 8	10 11	14 15	21 22	30 31	44 45	↑	↑	↑	↑	↑	↑	↑	↑	↑	↑	↑
P	800	↓	↓	↓	0 1	1 2	2 3	3 4	5 6	7 8	10 11	14 15	21 22	30 31	44 45	↑	↑	↑	↑	↑	↑	↑	↑	↑	↑	↑	↑
Q	1250	↓	↓	0 1	1 2	2 3	3 4	5 6	7 8	10 11	14 15	21 22	30 31	44 45	↑	↑	↑	↑	↑	↑	↑	↑	↑	↑	↑	↑	↑
R	2000	↓	0 1	1 2	2 3	3 4	5 6	7 8	10 11	14 15	21 22	30 31	44 45	↑	↑	↑	↑	↑	↑	↑	↑	↑	↑	↑	↑	↑	↑

⇩ = Use first sampling plan below arrow. If sample size equals, or exceeds, lot batch size, do 100% inspection.
⇧ = Use first sampling plan above arrow.
Ac = Acceptance number
Re = Rejection number

BS 6001—Table II.

Fig. 2. Single sampling plans for normal inspection (Master table).

COMPANY B

QUALITY PROCEDURE QPB 002

CONFIGURATION AND CHANGE CONTROL

Compiled by...

Checked by...

Approved by...

Date...

CHANGE HISTORY PAGE

Document status	Date issued	Number of pages	Changed pages	Change/ defect no.
XX	XX	XX	XX	XX

CONTENTS LIST

1 INTRODUCTION

1.1 PURPOSE

To ensure both hardware and software integrity and the continuity of the overall project, there is a need to establish a configuration management approach which must be followed through the project life-cycle.

1.2 SCOPE

All hardware, software, firmware and documentation produced by Company B or otherwise as dictated by the contract.

1.3 TERMS AND ABBREVIATIONS

CI—configuration item: the smallest uniquely identifiable item which when combined with other items gives units and modules.

CM—configuration management: a method to control updates and changes to CIs at critical points in the life-cycle.

CCP—configuration change panel: the CCP evaluates and authorises changes to controlled items.

CFA—configuration functional audit: audit of a CI's function against its requirements.

CPA—configuration physical audit of the 'as built' CI.

2 APPLICABLE DOCUMENTS AND REFERENCES

QPS 003—Design and Review
QPS 004—Software Development and Standards
QPB 008—Documentation Format and Standards
QPB 019—Internal Quality Audits
WIS XXX—Engineering Drawing
Drawing Register

3 HARDWARE CONFIGURATION MANAGEMENT

A hardware configuration management (CM) programme is needed that establishes baselines at critical points in the hardware development life-cycle. The configuration identification of the CI forms a basis or framework for establishing control of the configuration and changes to it.

To establish a baseline, identify the items that are to be included and then proceed by controlling changes to them (configuration control), monitoring the changes (configuration auditing) and reporting the status of the baseline as it changes (configuration status accounting).

This is done until another critical point in the development life-cycle is reached. A new baseline is then established for controlling changes to the CI.

A subset of hardware CM is data management. Data management controls hardware documents, including fabrication and assembly drawings; maintains error/change forms; and provides a basis for ensuring the hardware documents are consistent with other project documents.

3.1 IDENTIFICATION

The hardware configuration is identified on the drawing register.

The following paragraphs describe each baseline and list the documents required to describe the hardware system at these critical points.

(i) *Requirements baseline*—this configuration identification is fixed at the hardware configuration item requirements review. It includes hardware requirement specifications and functional interface descriptions.

(ii) *Development baseline*—this configuration is developed at the critical hardware design review and consists of detailed design specifications and detailed functional interface specifications.

(iii) *Product baseline*—this establishes the configuration of the hardware after testing, prior to release. The product baseline consists of updated development baseline documentation.

(iv) The *document master index*, shown in Fig. 1, is used to summarise the documents and their issues which make up a product at any particular baseline.

3.2 CONTROL

After a configuration is established, controls are exercised to prevent any unauthorised changes to the configuration. The primary control is the change control panel (CCP). This requires that all hardware design changes are documented, evaluated and coordinated by being submitted to the CCP for approval.

AB/089/991
ISSUE 6.0

		AB/089	BL1	BL2	BL3	BL4	BL5	BL6
Requirements Spec		100	1·0	2·0	2·0	2·0	2·0	2·0
Functional Spec		200	1·0	2·0	2·1	2·1	2·1	2·1
Hardware Tech Spec		300		1·0	2·0	2·0	2·0	2·0
Software Tech Spec		400		1·0	2·0	2·0	2·0	2·0
Subsystems—	SUPYS	410				1·0	2·0	2·0
	INIT	420				1·0	2·0	2·0
	FULLCHECK	430				1·0	2·0	2·0
	DOFOREVER	440				1·0	2·0	2·0
Modules—	ADDRESS	441				1·0	2·0	2·0
	DETDIAG	442				1·0	2·0	2·0
	DETFIRE	443				1·0	2·0	2·0
	CLOCK	444				1·0	2·0	2·0
	DISPSTAT	445				1·0	2·0	2·0
	CHKKEY	446				1·0	2·0	2·0
	REPORT	447				1·0	2·0	2·0
Quality Plan		900	1·0	2·0	2·0	2·0	2·0	2·0
TestSpecs—	MOTHER Bd	911					1·0	2·0
	CPU Bd	912					1·0	2·0
	I/O	913					1·0	2·0
	COMMS	914					1·0	2·0
	PSU	915					1·0	2·0
	I/O CPU	921					1·0	2·0
	COMMS/CPU	922					1·0	2·0
	FUNCTIONAL	931					1·0	2·0
	I/O LOAD	932					1·0	2·0
	MARGINAL	933					1·0	1·0
	MISUSE	934					1·0	2·0
	ENVIRONMENT	935					1·0	2·0
EPROM—XYZ							1·0	—
PROM—ABC							—	1·0

Fig. 1. Document master index.

Change requests to the CCP must contain the following information:

—description of hardware change
—impact on project
—cost of change
—effects on specifications, drawings and interfaces
—justification and reason for change
—schedule impacts
—impacts on safety, reliability, QA, testing, etc.

Three configuration change levels need to be defined which classify changes to baselined items. Each level requires different processing and reviewing procedures.

Level 1 changes: major modifications to a product as a result of a requirements change and requires review and approval by the project and client prior to implementation.

Level 2 changes: minor low-risk changes, or documentation updates requiring CCP authorisation and customer agreement.

Level 3 changes: internal changes, or modifications to CIs or documents still under internal control that have not been released to level 1 or level 2 control.

Figure 2 shows the change request form used to control changes.

3.3 AUDITS

Audits are performed to ensure that items under CM agree with the established baselines.

To verify that the CI product baseline agrees with the design specifications, a configuration functional audit (CFA) and a configuration physical audit (CPA) are carried out.

The FCA is a means of validating that development of a CI has been completed satisfactorily; it is a formal prerequisite to the PCA. The following list describes the FCA process:

—Review test/analysis results to ensure that testing is adequate, properly performed and, if applicable, certified by the customer.
—Review test/analysis results to verify that actual performance of the CI complies with its development or specifications and that sufficient test results are available to ensure the CI will perform in its hardware environment.

252

Date   Project   Project No.:   Change No.:

Subject/Item:

Raised by:

Details of change:

Copy to Engineering and Quality Manager

Analysis of change by CCP:

Cost: Schedule:

Documentation: Reliability:

Testing: Safety:

* Agreed CCP   (signature of reviewers)
* Not agreed CCP (signature of reviewers)......................................

(* Delete as applicable)

Date ....

Copy to Engineering and Quality Manager

Fig. 2. Change request form.

—Review all error/change/concession forms to customers' specifications and standards. This is to determine the extent to which the equipment undergoing FCA varies from applicable specifications and standards, and to form a basis for satisfactory compliance therewith.

The PCA is a means of establishing the product configuration identification, which was used initially for the production and acceptance of units of a CI. It ensures that

(1) the as-built configuration of a CI matches the product documentation, and
(2) the acceptance testing requirements prescribed by the documentation are adequate for acceptance of production units of the CI.

The following activities are carried out to ensure a successful PCA:

—Compare drawings with equipment to ensure that the latest drawing change numbers have been incorporated into the equipment, that part numbers agree with drawings, and that the drawings are complete and describe the equipment accurately. Review parts list to ensure client approval of non-standard parts.
—Verify that acceptance test results have been reviewed to ensure that testing is adequate, properly performed and, if necessary, agreed with the client.
—Verify that shortages and unincorporated design changes are listed in a report and have been reviewed.
—Examine the report to ensure that it defines the equipment adequately and that unaccomplished tasks are included as deficiencies.
—List all concessions to customer specifications and standards that have been approved. This forms the basis for satisfactory compliance with the customer's specifications and standards.
—Review the engineering release and change control procedures to ensure that they are adequate to control the processing and formal release of engineering drawing changes.

3.4 CONFIGURATION STATUS ACCOUNTING

Configuration status accounting is needed at the time the product's configuration identification is approved/accepted. Ensure that configuration status accounting is maintained, normally until the last unit of the configuration type, model or series is delivered.

Documentation must be established that, as a minimum, includes identification of the following:

—essential hardware configuration item and data elements;
—contractual information required in the records/reports for each configuration item;
—proposed level 1 changes to configuration and the status of such changes;
—approved changes to configuration, including the specific number and kind of items to which these changes apply, the implementation status of such changes, and the personnel responsible for their implementation.

3.5 DATA MANAGEMENT

Approved fabrication and assembly drawings are placed under data management control. Retain these drawings in a data control area. For full details see WIS XXX (Engineering Drawing).

4 SOFTWARE CONFIGURATION MANAGEMENT

The details given for hardware configuration management apply equally to software. Software items to be controlled are given below.

4.1 REQUIREMENTS DEFINITION

The main items of the software requirements definition are a software requirements specification and a database design specification. The requirements specification defines in detail the criteria (e.g. functional, performance, interface) to be used to assess the acceptability of the software product, to be designed and implemented.

The database design specification establishes the design of the database, including the physical content and structure. The software requirement specification is the requirements baseline.

This baseline is reviewed with, and may be formally approved by, the client. Each element of the specification must be reviewed or inspected by a peer. At the conclusion of build and acceptance testing, the specifications contents are audited for consistency with approved changes.

4.2 DESIGN PRODUCTS

The main items from the design process are the software acceptance test plan, any database details and the software design specifications.

The software acceptance test plan identifies the tests that will be conducted to demonstrate to the client that the software product meets all the requirements defined in its requirements baseline. This test plan is normally reviewed with the client.

Although a test plan is not usually part of a formal baseline, it should nonetheless be approved by the client. The database subschema describes all the items in the database and specifies the interrelationships among the elements of the database.

The software design specification defines the software design to be implemented. A preliminary design is developed early in the design process and is reviewed with the client during a preliminary design review. The preliminary design evolves into the final design during the remainder of the design process and may be reviewed with a client during a critical design review.

After holding a review, update the design specifications and test plans to correct problems detected during the review. When the client agrees that the updated specification describes a product that will meet its requirements, establish the design specification as the development baseline.

Each design diagram defining the preliminary and idealised software and database designs is subject to inspection by a peer.

The diagrams and final design specification need not be individually inspected. Internal walkthroughs by teams of peers and engineers usually suffice for the final design if each preliminary and idealised design diagram has been inspected.

During implementation and testing, the design specification is likely to change. At the end of build and acceptance testing, the specification is checked for consistency with the tested software. All discrepancies are reported as problem reports to be resolved in future builds or releases. The audited design specification is the principal element of the product baseline and the foundation for the development baseline of the next build or release.

4.3 PRODUCT IMPLEMENTATION

Each item is subject to an internal peer inspection. The inspections of unit designs and test plans should be formal. All other inspections (e.g. unit code) need be less formal but should be documented one-to-one inspections between authors and peers. Unit designs and code listings become part of the product baseline at the end of build testing.

All products of software implementation are subject to quality audit to ensure their completeness and conformance to standards upon their release for build testing. Audit and software build testing are allowed to proceed in parallel for large developments but require both to be completed before certifying the CI as complete.

Any products found by the audit to contain major deficiencies are returned to the implementation group for correction and redelivery. Change/defect forms are raised.

4.4 TESTING PRODUCTS

Before starting a software build test, the procedure for running the test has to be prepared. Before reporting a software build test as complete, the results of the test are recorded and the system engineer and quality need to inspect the summary. Software build testing is audited against the procedures and summary test reports.

Before starting acceptance testing, the procedures for running the tests are prepared. As procedures are prepared, they must be subject to inspection by a system engineer and review by the client. Acceptance tests must be delayed until the client agrees that the test procedure is adequate. Before reporting an acceptance test as complete, the results of the test must be prepared as a report.

4.5 SOFTWARE REVIEWS

Like hardware reviews, software reviews can involve requirements, preliminary design and critical design (see procedure QPS 003).

The section on software will not be detailed any further since Company B's overall policy on software and hardware has been discussed above.

5 ERROR/CHANGE REPORTING

If an item is controlled (i.e. under configuration management) and is found to contain an error, to be defective or to need changing, then a change request shall be raised (Fig. 2).

Requests will be assigned a serial number so that each is uniquely identified. A serial register will be maintained by the engineering and quality manager, from whom successive numbers will be obtained.

The change request shall be cross-referenced to any error/concession/ defect sheets and changes will be assessed according to their change levels (see Section 3.2).

COMPANY B
QUALITY PROCEDURE QPB 004
SOFTWARE DEVELOPMENT AND STANDARDS

Compiled by...

Checked by...

Approved by...

Date...

CHANGE HISTORY PAGE

Document status	Date issued	Number of pages	Changed pages	Change/ defect no.
XX	XX	XX	XX	XX

CONTENTS LIST

1 INTRODUCTION

1.1 PURPOSE

No one specific design technique is called for. This procedure provides general methods for a design methodology. This is considered desirable owing to the abundance of design methods and the fact that the method chosen is often highly project-specific.

Software design techniques enable the requirements to be addressed in a structured manner, linking the stages of the design process, and their evaluation criteria.

Where possible all the terminology in this procedure is generic, owing to the large number of design methodologies.

The above comments also apply to the programming principles and practices described.

1.2 SCOPE

This standard applies to all software design and programming carried out for or on behalf of Company B except where contract conditions state otherwise.

This document refers equally to firmware and software.

1.3 TERMS AND ABBREVIATIONS

Design module—a generic term used to mean a single compilable or assembled design module which is mapped to a software coded module.

Diagram—a generic term used to mean a graphical representation either in requirements or design documentation.

Module—a generic term used to mean a single discretely identifiable part of a system or software.

Software coded module—a generic term used to mean a single procedure or sub-routine derived from a design module and performing a discrete function in software.

Top-down—the structure imposed on a system through breaking it down into component parts or levels: system, sub-system, task or module. Applies to both design and software units.

2 APPLICABLE DOCUMENTS AND REFERENCES

Yourdon, E. & Constantine, L., *Structured Design*. Yourdon Press, 1978.

Jackson, M. A., *Principles of Program Design*. Academic Press, London, 1975.

Myers, G. J., *Reliable Software Through Composite Design*. Van Nostrand Reinhold, Wokingham, UK, 1975.

Gane, C. & Sarson, T., *Structured System Analysis: Tools and Techniques*. Prentice-Hall, Hemel Hempstead, UK, 1979.

Page-Jones, M., *Practical Guide to Structured Systems Design*. Prentice-Hall, Hemel Hempstead, UK, 1988.

Smith, D. J. & Wood, K. B., *Engineering Quality Software*, 2nd edn. Elsevier, London, 1989.

QPB 008—Documentation Format and Standard

QPS 003—Design and Review

QPB 002—Configuration and Change Control

3 FEATURES OF GOOD DESIGN TECHNIQUES

Company B's approach is not restricted to any one design technique, since these are often project-specific. However, the technique used should meet certain criteria.

The most important criterion is that the design technique should complement the requirements definition and should define explicit methods for deriving an initial (preliminary) software design configuration from the requirements, including its diagrams or graphical representations.

The selected technique should also:

(a) make it easy to cross-reference requirements with the design;
(b) provide guidance on how to evolve from a preliminary to an idealised design;
(c) provide guidance on how to detect and correct design features that might reduce software maintainability;
(d) provide intermediate documents that can be used to measure progress;
(e) define criteria for certifying the intermediate products.

All graphical representations or diagrams should:

(a) show the software product as a top-down set of configuration diagrams;

(b) portray any database as a top-down set of diagrams showing the hierarchical relationships among all sub-schema;

(c) define the function of each software module;

(d) identify data and control interfaces between modules;

(e) specify files and global data accessed by each module;

(f) include enough information to understand and evaluate the design without reference to the internal logic of software coded modules;

(g) permit easy updates for requirements and design changes.

The sources of some design techniques are listed in Section 2.

3.1 ORGANISING A DESIGN SPECIFICATION

In assembling a software design specification, arrange the design top-down to show a smooth progression from the highest to the lowest level.

4 THE SOFTWARE DESIGN PROCESS

The Company B design process is structured to ensure that the software is implementable, maintainable and complies with requirements.

It recognises that the first design generated, although sometimes mediocre, is a sound starting point for a good final design. The design evolves in three stages:

(a) Preliminary design—an initial (first-cut) configuration that explicitly accounts for all functions and data shown in the requirements and on requirements diagrams.

(b) Idealised design—an expanded version of the initial configuration that accounts for all detailed processing and interface constraints, and which is optimised for product maintainability.

(c) Final design—the idealised design, modified to account for performance requirements and resource constraints.

4.1 DESIGN EVOLUTION

The three-step evolutionary design process depends on reducing the scope of the preliminary design effort so that it can be done quickly. Company B will use a design technique that makes it easy to derive explicitly a preliminary configuration from the requirements and its diagrams.

The preliminary design need not meet all the requirements. It should, however, provide all the information needed to determine whether the software will perform all the functions and process all the data as called for in the requirements.

The preliminary configuration is expanded to show each major function to be performed in software. Idealise the expanded configuration to minimise features that would make the software difficult to implement, test and maintain. Existing software to be reused is identified.

Finally, modify the idealised configuration, as needed to use or interface with existing software and to comply with explicit performance requirements called for in the system or software requirements specification.

4.2 SYSTEM AND SUB-SYSTEM DESIGN PRODUCTS

4.2.1 Design Product Contents

Present the design primarily in the form of configuration design diagrams to show:

(a) the hierarchy of control among the design modules comprising the software product;

(b) the purpose of each design module; and

(c) the data and control interfaces among them.

A design module is a generic term used to mean a single item to a compiler or assembler. In the idealised and final designs, it should represent an item that will contain less than perhaps 100 executable source statements when coded. Define all data and control interfaces shown on the configuration design diagrams in a data dictionary. Each diagram is subject to peer inspection and design review and audit, thus ensuring consistent design quality.

5 CRITERIA FOR EVALUATING SOFTWARE DESIGNS

5.1 GENERAL

Company B will endeavour to design easily maintainable software such that both implementation and testing will be straightforward. Company B will endeavour to make correlating a design with requirements easy, so that deviations will be readily detected before they severely impact on costs and schedules.

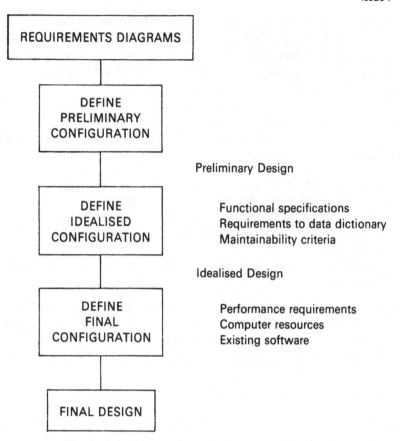

Fig. 1. Criteria to evaluate design configurations.

Company **B** will use different criteria to evaluate preliminary, idealised and final designs. Figure 1 identifies the principal criteria used to review the designs produced during each step in the design process. The three-step process will result in a design representing a highly maintainable product that meets functional and performance requirements, complies with design constraints, and is easy to implement, test and modify.

5.2 PRELIMINARY DESIGN CRITERIA

A preliminary configuration that explicitly accounts for all information shown on the requirements diagrams is to be defined. A process of iteration, as detailed below, then follows.

265

Identify at least one design module for each function and data element shown on the requirements diagrams. State the design modules' functions as simple imperative sentences using terms that appear on the requirements diagrams. Include on the module interfaces only control items (e.g. flags) that would be noticed by their absence.

Keep all design modules that are sensitive to the format, source and destination of external data at the bottom of the configuration.

If possible delay until the third step (the final design step) making any packaging decisions, such as:

—naming design modules;
—grouping design modules into tasks or sub-systems;
—differentiating between large and small design modules;
—indicating that some design modules will be embedded in other modules;
—combining design diagrams or similar design modules across diagrams.

By following the above strategy Company B's designers will be able to quickly define a preliminary design that clearly provides all functions and processes for all the data shown on the requirements diagrams. It will be a solid foundation for a quality design that meets all requirements.

5.3 IDEALISED DESIGN CRITERIA

The preliminary design is expanded to implement each major action in software coded modules.

The design is detailed to a level such that every module can be coded in less than 100 executable source statements in a high-level coding language. Identify a software coded module for each step needed to meet each function and data requirement.

Delay combining several trivial steps into single design modules until early in the final design effort, even if the trivial steps seem to represent very small design modules (e.g. modules that may require less than 25 executable source statements).

The design module interfaces are expanded to include all control items needed for design validity (e.g. error and status indicators). Each design module interface is defined in a design data dictionary.

Include in the design data dictionary all data items defined in the requirements data dictionary. Combine similar design modules across design diagrams. All design modules are named. Preserve, as much as

266

possible, the terms used in the preliminary design diagrams. Make sure all diagrams are consistent with each other.

Clutter is reduced by excluding standard functions provided by the high-level language (e.g. SORT, SIN, COS) and simple functions provided by operating systems (e.g. OPEN, CLOSE, READ, WRITE).

All dependencies between modules and physical data structures (e.g. files and global tables) are shown. Avoid defining a module that needs both data residing in a file or global table and depends on the physical structure or location of the data.

5.4 FINAL DESIGN CRITERIA

The idealised design is modified, as needed, to meet performance requirements, resource constraints (e.g. memory) and other design constraints. The modified design is used as the basis for packaging modules into suites of software coded modules.

The design is changed, if necessary, to account for asynchronous processing and for batch versus on-line functions. Identify all modules that will be embedded in other modules and which will not be coded separately. Update the design data dictionary to reflect all module interfaces changed while generating the final design.

6 PROGRAMMING PRACTICES AND PRINCIPLES

Although there is little agreement over specific software coding practices and principles, there is general agreement that code should be produced with personnel in mind. Code is produced for other people and not for machines.

Thus the aim of Company B's programming practices and principles is to produce well-structured and well-commented code which is easy to understand, unambiguous and therefore maintainable.

6.1 STRUCTURE

The following points must be considered when producing well-structured code:

(a)   A structured design for the program derived from the module design should be produced in a top-down manner, exhibiting its decomposition into a number of procedures/sub-routines.

(b) Each procedure/sub-routine should perform a separate well-defined function.

(c) Related procedures/sub-routines should be grouped together at a position that is advantageous to the addressing mode of the program sections that will use them. If indirect addressing is employed then they should be grouped together either at the beginning or end of the program unit.

(d) Common utility routines should be grouped together in separate program units, or in a general-purpose library.

(e) Procedure/sub-routine bodies should not be unduly long. No absolute limit is imposed, but one or two listing pages is suggested as a practical maximum.

(f) The use of unconditional branch instructions is not prohibited, but care should be taken to ensure that the modular structure is maintained.

(g) All sub-routines should be closed, i.e. data should only be passed in and out via specified data areas or registers/parameters, and where practicable should only have one entry and one exit point.

(h) All 'external' declarations must be explicitly written in the code at the beginning of the program, or contained in a called 'library' file.

6.2 READABLE LISTINGS

The following points must be considered to ensure that readable listings are produced:

(1) Normally each statement should appear in a new line. When assembler is used every effort should be made to ensure that statement columns do in fact line up.

(2) When block-structured languages are used indentation should be clearly used to show the scope of blocks and compound statements.

(3) Empty lines between statements should be used as punctuation to separate actions and highlight sub-functions or alternatives.

(4) Any functional break within a program unit (e.g. segment, main program loop, date areas) should be indicated by a line of asterisks.

(5) All identifier names should be meaningful, within the constraints imposed by the particular language, and should utilise the maximum number of characters allowed.

(6) Program layout and program/sub-routine headers are detailed in QPB 008.

(7) When available macro, or equates, facilities should be used to enhance the readability of code, particularly when referencing data items.

(8) Local macros, or equates, can be defined within a program, but those with wider applicability should be placed in a general macro/ equates 'library'.

(9) Comments should be used throughout the program to aid understandability. Such comments should enhance and emphasise the program structure rather than just reflect the operation of individual statements. Comments may also be used to highlight where modifications have been made.

(10) Data and variable declarations should always have an associated comment that indicates usage and range of values of the data.

(11) When using assembler languages the following points should be noted:

—Program transfers should only be made to labelled instructions. Constructs labelled $\pm$ should not be used.

—Values, character codes, etc., that are often used within a program should be declared as constants, or macros, rather than provided numerically for each usage.

—Instructions should only be used by reference to their mnemonics. Constructs should not be used to form data words that are interpreted as instructions.

—Code should not be self-modifying.

(12) All programs, whether produced in high level or assembler, should endeavour to use standard mechanisms when interfacing with the particular operating system being used. Any deviations should be highlighted and documented.

6.3 UTILISATION OF COMMON CODE

The following must be considered:

(a) From any specific requirement for a procedure/sub-routine, attempt to extract those aspects which are more general purpose in nature and hence might be performed by a library routine. This should be done whether or not a suitable library procedure/sub-routine exists.

(b) Check whether an appropriate routine exists. If so, use it, otherwise write one in a general-purpose manner. Note that, in the latter case, comments and data names should be appropriate to the general case.

(c) Library routines produced for one application will invariably require to be altered slightly, as new applications come to light with different requirements. This is acceptable *provided* all users are consulted about such changes and make corresponding changes in their method of use, if necessary. It is most important that no program exists in binary form using a library routine for which the source code is no longer available.

(d) Excessive generality of library routines is inefficient in both run-time, quantity of code to set up parameters and adapt to specific cases. It is likely that several procedures will exist to perform slightly different functions.

(e) Care should be taken to ensure that any potential problems with re-entrancy, routine nesting or data area requirements are carefully evaluated from a system viewpoint.

(f) Wherever possible library routines should be written in the highest level language available to the programmer.

6.4 TESTING/FAULT TOLERANCE

The following points should be considered:

(i) All programs should have a test strategy designed in parallel with the program itself. Adherance to the guidelines previously described in this document will result in a program which is significantly easier to test than one which has been produced in an ad-hoc manner.

(ii) A certain level of fault tolerance should be built into all programs so that no input, or output, to/from a program, or sub-routine, will cause the system to crash.

(iii) Each variable should be used for a single purpose only. The only exception to this is a temporary, or workspace, variable. This variable, whose name should indicate its temporary/workspace nature, may be used for different purposes provided each individual use is confined to a small area of program (i.e. such variables should not be set in one part of a program to be read in another).

(iv) Any monitoring facilities built into the program for testing purposes should be done in such a manner that the need for retrospective editing is minimised.

(v) It is usually necessary for appropriate recovery action to be built into real time systems. Certain key routines, therefore, should be able to determine their success or failure, in carrying out their specific function, and report this to the calling environment to enable recovery action to be taken at the appropriate level. It is not acceptable to jump directly from a routine to an error label, since this fails to provide a formal return path to continue normal operation after recovery has been effected.

6.5 PROGRAM LAYOUT

6.5.1 General Program Layout

The schematic shown below indicates an idealised layout of the program in terms of its constituent parts. Obviously sections may be added/deleted as necessary and the actual order manipulated to suit a particular language's requirements.

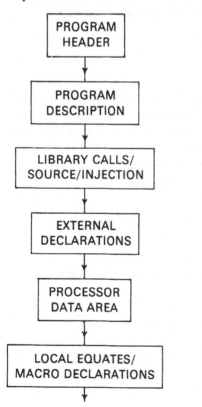

PROGRAM HEADER — Standard program header: this should be standardised wherever possible.

PROGRAM DESCRIPTION — Optional

LIBRARY CALLS/ SOURCE/INJECTION — Calls to standard library functions to be used by the program.

EXTERNAL DECLARATIONS — Declarations for routines/labels external to this program and those in this program, that may be used by other programs.

PROCESSOR DATA AREA — Operating system dependent data necessary to the operation of the program.

LOCAL EQUATES/ MACRO DECLARATIONS — Local declaration of equates and/or macro-definitions to be used by this program.

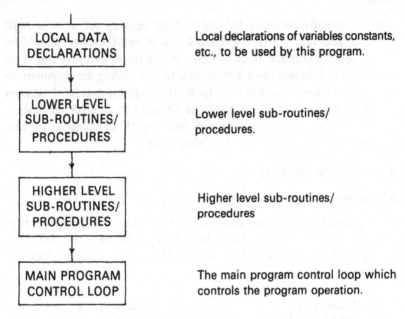

LOCAL DATA DECLARATIONS	Local declarations of variables constants, etc., to be used by this program.
LOWER LEVEL SUB-ROUTINES/ PROCEDURES	Lower level sub-routines/ procedures.
HIGHER LEVEL SUB-ROUTINES/ PROCEDURES	Higher level sub-routines/ procedures
MAIN PROGRAM CONTROL LOOP	The main program control loop which controls the program operation.

6.6 PROGRAM MODIFICATIONS

On any project program modification should be handled with great care to ensure that project status and testability is strictly maintained.

Normally pre-controlled software will be the responsibility of the team member allocated to the task, and any modification will only affect his particular work area. However, once software reaches a controlled status modifications must be assessed on a project-wide basis. Normally the following points should be considered:

(a) Identification of the need for modification (e.g. specification change, fault identification, etc.).

(b) Instigation of change/fault reporting mechanism.

(c) Implementation of modification after change/fault approval:
—update of program code;
—informal re-test program;
—update of program header, file name and associated information;
—highlight of code modification areas, if necessary.

(d) Formal re-test of software, if required, and archiving of new software.

COMPANY B
QUALITY PROCEDURE QPB 008
DOCUMENTATION FORMAT AND STANDARD

Compiled by...

Checked by...

Approved by..

Date...

CHANGE HISTORY PAGE

Document status	Date issued	Number of pages	Changed pages	Change/ defect no.
XX	XX	XX	XX	XX

CONTENTS LIST

8   SOFTWARE MODULE SPECIFICATION
   8.1   Functional Review
   8.2   Position of Module in Sub-system
   8.3   External Interfaces
   8.4   Module Design
   8.5   Module Description

9   FIRMWARE MODULE TEST SPECIFICATION
   9.1   Summary
   9.2   Interfaces
   9.3   Description

10   FIRMWARE MODULE TEST SCHEDULE
   10.1   Configuration
   10.2   Operation

11   PRODUCTION MANUAL
   11.1   Production Data
   11.2   Configuration Data
   11.3   Production Test Specification
   11.4   Production Test Schedule
   11.5   Test and Calibration Certificate

12   USER MANUAL
   12.1   Configuration
   12.2   Operation
   12.3   Maintenance
   12.4   Fault Reporting

13   SERVICE MANUAL
   13.1   Field Service
   13.2   Repair Service
   13.3   Other Relevant Data
   13.4   Description

Fig. 1   Hardware Document Structure

APPENDIX 1: SOFTWARE CONTENT GUIDELINES

APPENDIX 2: FIRMWARE FORMAT AND CONTENT

1 INTRODUCTION

1.1 PURPOSE

This document provides the procedures to enable hardware design documentation under development at Company B to meet the requirements of BS 5750 (Part 1) 1987.

1.2 SCOPE

This document defines the structure and format of hardware-related documents, and thus concerns itself with those documents below the system design level.

1.3 TERMS AND ABBREVIATIONS

None.

2 APPLICABLE DOCUMENTS AND REFERENCES

QPS 016—Calibration of Measuring Equipment
QPB 002—Configuration and Change Control
QPS 003—Design and Review
QPB 004—Software Development and Standards

3 DOCUMENT STRUCTURE

The documents that may be required from hardware development and the hierarchy of these documents is shown in Fig. 1. The base terms of reference for development are either a system design specification, sub-system design specification, user requirement or functional specification.

Although the documents defined in the following sections are shown as independent, the titles may be amalgamated into a common document or even dispensed with if irrelevant. For example, if no processor is incorporated into a design then the firmware documents are redundant.

Conversely, if the application warrants expansion a title may be divided among a number of documents. For example, the user manual may be divided into an operator's manual and a maintenance manual. These combinations are project-dependent.

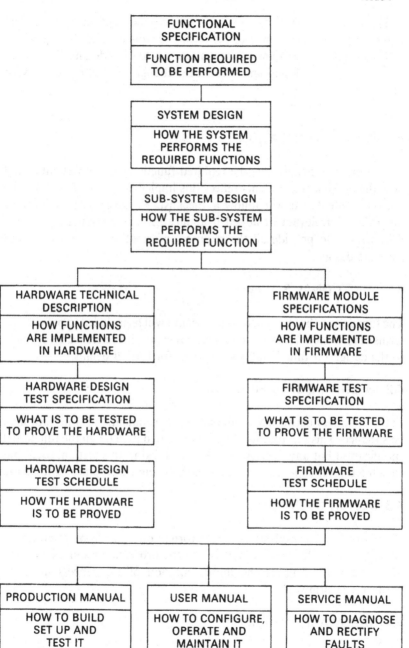

Fig. 1. Hardware document structure.

The following sections describe the contents of each title. The descriptions are constrained to the technical content of each document.

The purpose of design test documentation is to provide a fixed reference against which the design and subsequent changes to that design may be proved.

4 DESIGN SPECIFICATION

This document defines how the required functions are implemented and how the document's contents reflect this implementation. Guidelines for the full contents list will follow a checklist. However, sections may be discarded if irrelevant or added if applicable. The principal aim of the document is to provide all the required information for hardware and software design.

4.1 FUNCTIONAL REVIEW

The functions to be implemented by the unit will be included to provide the terms of reference. This should include a review, with reference being made to the design (or functional) specification for further details.

4.2 INTERFACE SPECIFICATION

In the boundary between the unit and its environment, each interface shall be specified. This information may be presented in detail in the functional specification but any deficiency will be rectified in this section, such that each interface is explained to enable independent development of the unit.

4.3 DESIGN OVERVIEW

This contains an analysis of the information flow from interface to interface for each function and defines the processing required en route with a view to defining the implementation of the process in either hardware or firmware.

4.4 HARDWARE OVERVIEW

From the design overview (Section 4.3), this section shall define the functions required of the hardware. At this point any hardware specifications, for example interface definitions that can be defined, may be specified.

4.5 FIRMWARE OVERVIEW

From the design overview (Section 4.3), this section shall define the functions required of the firmware, culminating in a firmware configuration, utilising accepted software design techniques, e.g. structured analysis and design. This section defines the individual modules and their interfaces.

5 HARDWARE TECHNICAL DESCRIPTION

This document describes how the functions assigned between the hardware blocks are to be implemented.

5.1 DESIGN OVERVIEW

This describes the hardware configuration by means of a block diagram and a functional description of each of the blocks and the interfaces between them.

5.2 BLOCK CIRCUIT DESCRIPTION

The operation of the circuit within each block is described.

5.3 TEMPERATURE AND ENVIRONMENTAL REQUIREMENTS

The environment and electrical characteristics under which the module or unit may be stored and operated are defined. The electrical and physical characteristics of the unit are specified.

6 HARDWARE DESIGN TEST SPECIFICATION

This document details what is to be tested to prove the suitability of the design to its application. The tests will be divided into electrical performance tests (Section 6.1) and environmental tests (Section 6.2), the latter being further divided into those tests that may be carried out at

Company B and the tests which require the specialist facilities of a test house.

6.1 ELECTRICAL TESTING

The tests to be carried out will depend upon the application. However, each unit will be tested to ensure that its particular function is provided by a design that complies with the storage and operating specifications, and procedures.

6.2 ENVIRONMENTAL TESTING

These tests will include temperature cycling and soak tests, static discharge, noise, drop, electromagnetic compatibility, vibration and humidity.
The testing will be divided into two stages:

(a) those tests that are carried out utilising Company B's equipment;
(b) those tests that require the use of the facilities of an independent test house.

Additional testing may be necessary to provide data for burn-in periods and reliability figures. This will take the form of accelerated life tests which will be defined during the development period.

7 HARDWARE DESIGN TEST SCHEDULE

This document details how the tests defined in the hardware design test specification are to be implemented.

7.1 CONFIGURATION

This section details the types of equipment required for the tests and defines the operational parameters that the equipment is to meet. Recommended equipment that meet these criteria is then listed.

7.2 SCHEDULES

For each test defined in the hardware test specification, a test schedule will be written explaining the test configuration, the input and the output conditions to be monitored, the acceptable range of the output conditions and the format to be used for recording the test results.

8 SOFTWARE MODULE SPECIFICATION

The document details how each software module defined in the sub-system design specification implements its functions.

8.1 FUNCTIONAL REVIEW

This section should list the functions required of the module, which needs to be reviewed.

8.2 POSITION OF MODULE IN SUB-SYSTEM

This section should describe, by diagrams, the relationship of the module to the rest of the sub-system.

8.3 EXTERNAL INTERFACES

This section lists and specifies in detail each of the module interfaces previously outlined in the sub-system design specification. It will also include details of both direct and indirect accesses to common data areas.

8.4 MODULE DESIGN

By use of suitable design aids (e.g. structured design) this section should describe the implementation of functions down to a sub-routine or procedure level, or to a coded block level of similar size. The section should also describe any local data areas used by the module.

8.5 MODULE DESCRIPTION

This section details the code structure of each sub-routine, procedure or block as defined in the module design section. These descriptions will be dependent upon the quality of the code, e.g. if the code is highly structured then pseudo-code is applicable. If, for whatever reason, the code is badly structured then flow charts may be more appropriate.

Whatever design aids are used to provide a cross-reference between document and listing, the test in the document should be the same as the header comment in the code.

For guidelines on software content and firmware format and content, see the appendices.

9 FIRMWARE MODULE TEST SPECIFICATION

This document details what is to be tested to prove that the module meets its design criteria. A document will be written for each independent task or module, whichever is the more convenient.

9.1 SUMMARY

This section gives an outline description of the objectives and scope of the tests.

9.2 INTERFACES

The module interfaces are defined such that sufficient information is available to enable the development of hardware and/or software required for the tests.

9.3 DESCRIPTION

This section describes the design of any hardware and/or software developed for the module tests.

10 FIRMWARE MODULE TEST SCHEDULE

This document details how the tests specified in the firmware test specification are to be carried out.

10.1 CONFIGURATION

This section details for each test the hardware and software required to carry out the test on the subject module. Any set-up conditions required of the hardware and software are to be specified.

10.2 OPERATION

This section lists for each test:

(a) the test objectives;

(b) procedures for conducting the test, including monitoring and recording requirements;

(c) predicted results and criteria for acceptability.

11 PRODUCTION MANUAL

This manual provides all the information required to enable the module or unit to be manufactured, configured and tested.

11.1 PRODUCTION DATA

This data will consist of the circuit diagram, parts list and artwork to enable the unit to be built to a module level. It will also contain any special constructional details that may have become apparent during development.

11.2 CONFIGURATION DATA

This data will consist of details which enable production to start producing the unit, i.e. linking information, programmable devices, gain settings, etc.

11.3 PRODUCTION TEST SPECIFICATION

This document will define the tests required to prove the correct operation of the unit.

11.3.1 Summary

This section gives an outline description and scope of the tests, which should include electrical, functional and environmental tests. The last of these utilises temperature cycling and elevated temperature soak tests.

11.3.2 Description

This section describes the design of any hardware and/or software developed for the production tests.

283

11.4 PRODUCTION TEST SCHEDULE

The document will define how the tests described in the production test specification are to be carried out.

11.4.1 Configuration

This section details the hardware and firmware (if any) required for each test and how the equipment is configured. The type of equipment is specified and the operational parameters required of the equipment defined. Equipment that fulfils these criteria is then specified.

11.4.2 Operation

This section details how the equipment is to be set up, the procedures for conducting the tests and the predicted results as a criteria for acceptability.

11.5 TEST AND CALIBRATION CERTIFICATE

For each production test schedule there will be a test and calibration certificate.

12 USER MANUAL

This manual explains to the user of the designed system how to configure it for its intended application, how to operate it, and how to provide routine preventative maintenance.

12.1 CONFIGURATION

This section details how the user sets up the system for the application and provides details of the environment for which it is intended. The specification may be useful here as it provides both electrical and environmental parameters.

12.2 OPERATION

This section explains how the system is operated in its intended application and the various symptoms that can occur due to faults.

12.3 MAINTENANCE

This section explains how an operator carries out maintenance procedures required to prolong reliable operation.

12.4 FAULT REPORTING

This section explains the procedure to be followed should a fault occur and provides a standard form for fault reporting.

13 SERVICE MANUAL

This manual details how faults may be diagnosed either from operational symptoms or defined test procedures, such that a course of action may be suggested to render the system operational, e.g. from replacing a fuse to replacing the unit.

13.1 FIELD SERVICE

This section defines procedures for the field service engineer to follow, such that a course of action may be arrived at to render the system operational in the shortest possible time. This course of action will probably consist of unit replacement.

13.2 REPAIR SERVICE

This section defines procedures for repair engineers to correct faulty equipment at the factory, and thus should enable the engineer to diagnose faults down to the component level.

13.3 OTHER RELEVANT DATA

This section contains information that will be required at some stage of servicing, e.g. circuit diagrams, parts lists, etc.

13.4 DESCRIPTION

This section describes the design of any hardware and/or software developed for the requirements of diagnosing faults.

APPENDIX 1: SOFTWARE CONTENT GUIDELINES

These guidelines are intended to enable the documentation to be related to the listings and are language-independent. The various comment formats are listed and their relationship to the documents explained. The actual formats are defined in Appendix 2.

(a) *File/module/program header:* if a 1:1 correspondence exists between file and module program then this comment should relate to the lowest level of breakdown defined in the firmware specification (see Appendix 2).

If a module/program consists of a number of files then the files required to create the module/program are defined in the file header (see example in ¡Appendix 2).

A file shall not contain more than one module.

(b) *Procedure/sub-routine/block:* this comment should relate to the lowest level of breakdown defined in the module design section of the module specification. (A block is defined as a code sequence comparable to a procedure or sub-routine but simple entry and exit are effected by commented 'goto' statements.)

(c) *Header comment:* this comment relates to the test used in the module description section of the module specification.

(d) *Code comment:* this comment is used alongside the code to give a better understanding of the listing. If the module description section is required to be documented to this detail then the test should be the same.

(e) *Change comment:* columns, e.g. 73–79, are reserved to flag line changes to the code and give the date of the change, e.g. 020 890.

APPENDIX 2: FIRMWARE FORMAT AND CONTENT

(a) *File header*
   * File Name
      —The complete file name as listed on the main disk directory.
   * Description
      —Description of the functions implemented by the contents of the file.
   * Author
      —Name of the engineer responsible for the maintenance of the file.

* File History
  —Who originally coded the file and when; list of changes to the
  file, who did the changes and when; change/defect number
  authorising the change.
Note: The list is maintained during development when the description
of the change will include function and routine affected. On release, at
Issue 1, the development history is deleted and all further changes are
referenced by configuration change control procedures.
* File Contents
  —A list of contents (sub-routines, procedures, code blocks, etc.)
  to form a directory of the structure:

Procedure name

* Included Files
  —List of file names that are included at compile time.
* Linked Files
  —List of file names required to be linked with this file to form a
  module or task.

(b) *Routine header*
* Name
  —Routine name.
Note: # is the number assigned to this routine in the file/module
contents list.
* Function
* Description of the Routines Function
* Input Conditions
  —Definition of conditions that are required for the correct
  operation of the routine.
* Output Conditions
  —Definition of conditions set up by the routine.
* Routines Called
  —List of routines called. An entry is of the form

Routine name   Function   A/X/L

(absolute/external/local)

COMPANY B

QUALITY PROCEDURE QPB 010

CONTROL OF PRODUCTION

Compiled by...

Checked by...

Approved by...

Date...

CHANGE HISTORY PAGE

Document status	Date issued	Number of pages	Changed pages	Change/ defect no.
XX	XX	XX	XX	XX

CONTENTS LIST

1 INTRODUCTION

1.1 PURPOSE

This procedure provides the methods to be used for identifying project and production items.

1.2 SCOPE

The procedure shall apply to all hardware items used in Company B.

1.3 TERMS AND ABBREVIATIONS

None.

2 APPLICABLE DOCUMENTS AND REFERENCES

QPS 009—Purchasing and Suppliers
QPS 016—Calibration of Measuring Equipment

3 ROUTING DOCUMENTATION

The method of identifying items and their quality and test status is by use of a routing documentation package (or production folder).

When each item is received at Company B a routing package will be raised by the engineering and quality department. It will consist of each of the forms shown in Fig. 1. Additional pages will be raised as required.

The document package, in a suitable envelope or file, will accompany the item at all times until customer acceptance. To achieve this the equipment serial number will be recorded on both the equipment itself and the form shown in Fig. 1.

After customer acceptance the package will be filed, by contract number, in the engineering and quality department. This procedure will also apply to purchaser-supplied equipment.

4 HANDLING/HOUSEKEEPING

The following will apply:

—Floors will be checked each morning and cleaned if necessary.
—Shelving will be cleaned monthly by vacuum.

Contract name... Number..

Date received...

Item...

Description...

    (a) Hardware
    (b) Software
        2.1 Description
        2.2 Media
        2.3 Issue/Status

Incoming Inspection

                      Quality Engineer...

                      Date ....

Defect Report Numbers

Change Requests Numbers

Scrapped .... Date .... Approved............................

Test History

Test procedure..   Test procedure..

Date of test/retest ....   Date of test/retest ....

Accepted ....   Accepted ....

Test procedure..   Test procedure..

Date of test/retest ....   Date of test/retest ....

Accepted ....   Accepted ....

Test completed ....   Date..

Customer acceptance ....   Date..

Notes:

Fig. 1. Routing package.

—Food and drink will not be consumed adjacent to equipment.

—Partly or fully assembled equipment will be covered when not being worked on.

—Shrouded ash trays will be provided in smoking areas away from the equipment.

—Benches are the responsibility of individuals who will keep their workplace tidy and clean.

—Personnel will be instructed in the principles of good housekeeping.

It is the engineering and quality manager's responsibility to maintain a schedule of responsibilities for these activities and to see that they are implemented.

COMPANY B
QUALITY PROCEDURE QPB 011
GOODS INWARDS AND INSPECTION

Compiled by...

Checked by...

Approved by ....

Date..

CHANGE HISTORY PAGE

Document status	Date issued	Number of pages	Changed pages	Change/ defect no.
XX	XX	XX	XX	XX

CONTENTS LIST

1 INTRODUCTION

1.1 PURPOSE

To describe the procedure for all goods received at Company B.

1.2 SCOPE

All items which arrive at Company B.

1.3 TERMS AND ABBREVIATIONS

None.

2 APPLICABLE DOCUMENTS AND REFERENCES

QPS 022—Statistical Techniques
QPS 012—Identification and Traceability
BS 6001—Sampling Procedures and Tables for Inspection by Attribute

3 PROCEDURE

All goods that arrive at Company B are received at goods inwards. The goods will be placed on the appropriate shelf in the goods inwards area to await inspection.

Bulk manufactured parts and electrical components coming into the goods inwards area are inspected in accordance with the appropriate quality sampling procedures in BS 6001.

All finished items, subassemblies and assemblies are inspected and 100% tested in accordance with appropriate drawings, specifications and work instructions, and are identified as follows:

Green Spot—Goods which have passed inspection.
Yellow Spot—Goods offered for concession since they do not meet requirements. These goods are kept on closed-off or locked shelves until the concession is resolved.

Red Spot—Non-conforming goods. These goods are normally
returned to the supplier with a letter but may be rectified
by agreement between the supplier and Company B.
During this period they are kept on closed-off or locked
shelves. Non-conforming goods include those where
items are less than on the purchase order.

COMPANY B
QUALITY PROCEDURE QPB 014
LINE INSPECTION

Compiled by...

Checked by...

Approved by...

Date...

CHANGE HISTORY PAGE

Document status	Date issued	Number of pages	Changed pages	Change/ defect no.
XX	XX	XX	XX	XX

CONTENTS LIST

11   NON-CONFORMING ITEMS ON PCBs AND ASSEMBLIES
11.1   Non-conforming PCBs
11.2   Non-conforming Assemblies

12   RECORDS AND LABELLING
12.1   Labelling and Marking
12.2   Recording

APPENDIX: CRIMP PULL-OFF FORCE

1 INTRODUCTION

1.1 PURPOSE

The purpose of this document is to lay down the procedure for inspection of unpopulated and populated printed circuit boards (PCBs), and panel main and subassemblies.

1.2 SCOPE

The document applies to the inspection of all PCBs and panel main and subassemblies produced by Company B.

1.3 TERMS AND ABBREVIATIONS

PCB—printed circuit board
BS—British Standard
mm—millimetres

2 APPLICABLE DOCUMENTS AND REFERENCES

QPB 011—Goods Inwards and Inspection
QPB 015—Test
QPB 017—Non-conformance, Corrective Action and Records
BS 5370—Guide to Printed Wiring (Design, Manufacture and Repair)
BS 5738

3 EQUIPMENT UTILISED

Continuity tester, bench magnifying glass, micrometer, vernier, screw-drivers and spanners.

4 MECHANICAL

4.1 INSPECTION OF UNPOPULATED PCBs

PCBs shall be checked for dimensions, including those of fixing holes, to the drawing. Warpage shall not exceed $\frac{1}{16}$ in per foot.

4.2 INSPECTION OF POPULATED PCBs

Components shall be located and fitted correctly. Special attention shall be given to correct fitting and positioning of transistors and integrated circuits.

Fixed resistor values shall be read from left to right and from bottom to top. Heat sinks shall be situated within the silk screen area as they may short out tracks.

Components shall be inspected for correct part numbers, values, tolerances and damage. There shall be no discolouring due to overheating.

Component leads shall be correctly formed to suit the pitch of the holes in the PCB so as to prevent stress being exerted on components. Ideally, wire ends should be bent at 45° in the direction of the printed track.

Component wire ends should be cropped to the correct height before soldering.

4.3 INSPECTION OF PANEL MAIN ASSEMBLY, INCLUDING SUBASSEMBLIES

4.3.1 Assembly

All assemblies and parts shall be inspected in accordance with the relevant drawings for dimensions, positioning and alignment.

4.3.2 Finished Metal Parts

All treated surfaces (i.e. painted or plated) shall be inspected for damage (scratches, etc.).

5 VISUAL

The PCBs shall be visually inspected:

(a) For cracks, burns, nodules and blisters.
(b) To establish that plated through holes are clean and free from any inclusions which may affect component insertion or solderability.
(c) To check that tinning of copper tracks is satisfactory on any exposed area used for soldering. There should be a smooth, bright, solder coating with limited traces of scattered imperfections such as pin holes.

(d) To check that silk screening is legible.

(e) To check that solder lands are free from solder resist.

5.1 WIRING

Wiring shall be inspected for correct routing and layout. It must be free from damage caused by sharp edges and be correctly secured by the use of cable ties wherever strain relief is necessary. Wire shall not obscure the space for module insertion. Where cable trunking is used, not more than 40% of the cross-sectional area shall be filled with wires.

All wire ends shall be correctly formed with sufficient insulation clearance and with no damage due to insulation stripping or soldering heat. In the case of multi-strand wires, all strands shall be present and included in the solder joint. Insulation run-back shall not exceed four times the diameter of the wire after soldering.

Where wrapped connections are specified, the number of bare wire turns shall not exceed five. They shall be secure and free from gaps and not overlap.

5.2 CRIMP CONNECTORS

Where crimp connectors are used they shall be inspected for correct forming and the pull-off force shall be tested in accordance with the Appendix.

6 SOLDERING

A satisfactory solder joint must conform to WI XXX.

7 ELECTRICAL

PCBs shall be checked for electrical insulation between power pins. In accordance with the wiring specification, all wire terminations shall be continuity tested for correct location.

7.1 EARTHLINKS

The resistance between the earthlink terminals and the chassis shall exceed XX MΩ.

302

8 REPAIRS TO PCB CONDUCTORS AND MODIFICATIONS

8.1 REPAIRS TO PCB CONDUCTORS

Damaged conductors may be bridged with bare copper wire, but if the damage is greater than 5 mm then the copper wire shall be sleeved. No more than two repairs are permitted per board. Alternative repairs must conform to BS 5370. Where solder resist has been removed, prior to repair, then it shall be restored whether by the original process or by the use of polyurethane lacquer.

8.2 MODIFICATIONS

There shall be no more than seven modifications to any board (i.e. track cuts or links). Where link wires are greater than 2 in (50·8 mm) long then they shall be secured to the board by a suitable method (i.e. adhesive bond) at points along the wire.

9 CLEANLINESS

After assembly and wiring, all PCBs shall be clean and free of flux and contamination, cropped ends and other loose particles.

10 SERIAL NUMBERS

All PCBs shall carry a serial number.

11 NON-CONFORMING PCBs AND ASSEMBLIES

11.1 NON-CONFORMING PCBs

In the event of a PCB being rejected then it shall be identified and segregated pending rework, replacement or return to supplier (in the case of a purchased board). A concession note (see Fig. 2 in QPB 017) shall be raised if applicable and the engineering and quality manager informed.

11.2 NON-CONFORMING ASSEMBLIES

In the event of rejection, all assemblies shall be segregated pending rework (see QPB 017).

12 RECORDS AND LABELLING

12.1 LABELLING AND MARKING

Where labelling and marking is specified (i.e. name plate labels, serial numbers, fuse ratings, warnings, etc.) it shall be checked for positioning and correctness. Hazard warning labels must be fitted, in accordance with BS 5378, to all equipment wherever there is a risk of electric shock.

12.2 RECORDING

Routing documents shall be kept for each batch of PCBs during all stages of manufacture.

APPENDIX

CRIMP PULL-OFF FORCE

Wire size (mm^2)	Force in Newtons
0·28	55·6
0·39	70·1
0·50	78
0·75	101
1·00	134
1·50	206
2·50	381
4·00	600
6·00	807

CRIMP COLOUR CODE

Crimp colour code	Wire size (mm^2)	Wire size (SWG)
Red	0·5–1·5	22–16
Blue	2·5	16–14
Yellow	4–6	12–10

COMPANY B
QUALITY PROCEDURE QPB 015
TEST

Compiled by...

Checked by...

Approved by...

Date...

CHANGE HISTORY PAGE

Document status	Date issued	Number of pages	Changed pages	Change/ defect no.
XX	XX	XX	XX	XX

CONTENTS LIST

1 INTRODUCTION

1.1 PURPOSE

The purpose of this procedure is to lay down the method for testing printed circuit boards and completed equipment.

1.2 SCOPE

In-house final test of PCBs.
In-house final test and client witnessing of completed equipment.

1.3 TERMS AND ABBREVIATIONS

PCB—printed circuit board
ATE—automatic test equipment
MOS—metal oxide semiconductor
PROM—programmable read-only memory

2 APPLICABLE DOCUMENTS AND REFERENCES

QPB 010—Control of Production
QPB 014—Line Inspection
QPS 012—Identification and Traceability
PCB and equipment test specifications—as appropriate

3 EQUIPMENT UTILISED

The equipment used in testing will include:

Heat soak tent
Thermocouples
Temperature chart recorder
Multimeter
24V power supply
Oscilloscope type *xxxx*
ATE type *xxxx*
Antistatic mats and wristbands

Test rigs
Test software **PROMS**
Burn-in oven

4 TESTING POPULATED PRINTED CIRCUIT BOARDS

After inspection, all PCBs shall be tested in accordance with their respective test specifications.

4.1 AUTOMATIC TEST EQUIPMENT

Wherever an ATE test program has been written the *xxxx* tester will be used for the PCBs. In cases where no ATE program has been written a full manual test procedure will be used and each PCB will be bench-tested.

4.2 BURN-IN

After the initial test, all PCBs shall be placed in suitable racks in the burn-in oven and subjected to a temperature cycle for a period of time as stated in the test specification.

4.3 RETESTS

After the completion of the burn-in cycle, all PCBs shall be retested as in Section 4.1. Where faults are found, the boards will be repaired and then subjected to a further burn-in (4.2) before retest. PCBs which fail a second or subsequent retest shall be referred to the engineering and quality manager.

5 TESTING COMPLETED EQUIPMENT

5.1 SETTING UP AND TESTING

All modules shall be placed in their appropriate rack/panel locations and the test equipment connected as indicated in the equipment test specification. All module and equipment serial numbers will be checked against the routing documentation or production folders (QPS 012). The equipment tests will then be carried out according to the equipment test specification and the results recorded.

5.2 HEAT SOAK

The equipment shall be placed in the heat soak tent and subjected to a temperature cycle in accordance with the functional and test specifications. Power shall be maintained to the equipment during the test, and the temperature shall be monitored, by means of thermocouples and the temperature chart recorder, at specific points on the equipment as detailed in the test specification.

During the heat soak, specific functional tests will be performed on the powered equipment as detailed in the test specification. The ambient temperature, within the heat soak tent, shall be monitored at hourly intervals for conformance to the temperature cycle requirement.

6 NON-CONFORMING ITEMS

Rejected items shall be clearly identified and segregated pending sentencing and/or rework.

7 RECORDS AND CERTIFICATES

The copy of the test specification shall be used as the test record and a positive record of each successful test, as well as any non-conforming results, shall be made on the test specification. At the completion of test the routing documentation shall also be marked to indicate the test status of the PCB or equipment.

For customer/client witnessed tests, on completion of the test, an acceptance certificate shall be raised by the engineering and quality manager and signed by himself or a delegated member of the quality department.

8 PRECAUTIONS FOR SENSITIVE DEVICES

Where static sensitive devices are handled, they shall be protected by using antistatic mats and wristbands.

APPENDIX: PRECAUTIONS WHEN USING MOS DEVICES

All MOS devices can be damaged by the high-intensity electric fields in the gate-oxide region. They can cause the gate-oxide, normally 1200 Å in thickness, to rupture. A conductive path is thus created between the metal, or silicon gate and a diffused region, rendering the device unusable. In view of the normal gate-oxide thickness, a potential difference of 60V should not be exceeded between the gate and any other of the device terminals if a reasonable safety margin is to be applied. Since gate capacitance is typically in the order of 10 pF a very small static charge will create a 60V potential difference. The following simple precautions are thus recommended:

(a) *General*
Antistatic clothing should be worn when handling MOS devices. Cotton is preferred to nylon. Table tops should be at ground potential with a conductive ground rail which should be touched before handling each device. The device leads should be at ground potential.

(b) *Assembly*
Devices should be placed on conductive trays so that all leads are electrically shorted together. Soldering irons used for soldering MOS devices should have rounded tips.

(c) *Testing*
Attempt to eliminate voltage transients from power supplies used during testing. Do not insert or remove MOS devices from the circuit whilst power is applied. To remove a device, touch the conductive rail first before grasping the device.

(d) *Storage and transport*
MOS devices should be transported in conductive trays. Metal or conductive rubber trays are preferred to aluminium- or foil-lined trays. MOS devices should not be placed in 'foam' or polystyrene-type trays.

COMPANY B

QUALITY PROCEDURE QPB 017

NON-CONFORMANCE, CORRECTIVE ACTION AND RECORDS

Compiled by...

Checked by..

Approved by ....

Date..

CHANGE HISTORY PAGE

Document status	Date issued	Number of pages	Changed pages	Change/ defect no.
XX	XX	XX	XX	XX

CONTENTS LIST

1 INTRODUCTION

1.1 PURPOSE

This procedure is intended to provide detail for corrective actions and records for defective items either produced by Company B or purchased from outside.

1.2 SCOPE

All defective and non-conforming hardware and software items.

1.3 TERMS AND ABBREVIATIONS

None.

2 APPLICABLE DOCUMENTS AND REFERENCES

QPB 002—Configuration and Change Control
QPS 003—Design and Review
QPB 004—Software Development and Standards
QPS 009—Purchasing and Suppliers
QPB 014—Line Inspection
QPB 015—Test
QPS 013—Stores
QPB 010—Control of Production
QPS 011—Goods Inwards and Inspection

3 CONTROL OF NON-CONFORMING ITEMS

This covers all defective items and all failures, including those related to software, during operation and test which must be formally recorded on a defect sheet (Fig. 1).

4 CORRECTIVE ACTIONS PROCESS

The engineering and quality manager has the responsibility and authority for instituting corrective action. For each reported problem he will

Date   Project   Project No.   Sheet No.

Subject/Item:

Raised by: Date:

Details of defect: Change request No.:

Copy to Engineering and Quality Manager:

Corrective actions:

Analysis of defect:

Cost: Schedule:

Documentation: Reliability:

Testing: Safety:

Error/change request No.:

Defect correctable:

Not correctable, concession needed:

Agreed Engineering and Quality Manager:

Signed ....Date ....

Fig. 1. Concession/defect sheet.

ensure:

(a) The impact of the problem will be investigated with respect to cost, schedule, performance, reliability, safety and client satisfaction.

(b) All potential causes of the problem will be reviewed, as will their effect on other aspects of the contract.

(c) When the cause has been established preventive action must be taken as appropriate to the magnitude and scope of the problem. This may involve amendments to:

—manufacturing processes
—software (including compilers, tools, etc.)
—quality procedures and instructions
—specifications

(d) The effect of preventive measures, once taken, will be monitored for their effectiveness.

(e) Where work in progress is affected, the appropriate routing or production documents will be amended. It is likely that existing items will be modified or in some way used, in which case a concession/defect sheet will be raised (Fig. 1). In some cases the sheet will be raised but subsequent investigation will lead to a concession. Where goods inwards items are involved the requirements of QPS 011 will apply.

(f) Permanent changes, resulting from corrective actions, will result in changes to the process or quality procedures.

5 NON-CONFORMING ITEMS

Non-conforming items must be clearly labelled and segregated from the work area. Hardware, software and documentation which fails to conform to requirements shall be quarantined pending sentencing. Access shall be controlled.

The decision, in respect of each item, may be one of the following:

—accept as is with a concession
—return to supplier for rework/redesign
—reject and reorder from another supplier
—rework/rectify in-house
—amend the order

5.1 CONCESSIONARY PROCEDURE

Concessions shall only be valid for the items stated and under the conditions specified.

Where a concession involves a supplier then the concession shall become a contractual document and agreement, in writing, will be established with the supplier.

5.2 SENTENCING

In the event of a failure or defect it will be necessary to sentence the item. In the first instance the decision to rework, raise a change or defect request, or scrap the item will be made by the engineering and quality manager. Where a change request has been raised, the number cross-reference will be added to the concession sheet.

5.3 CONCESSION AND DEFECT ANALYSIS

A monthly summary of the preceding month's records will be prepared by the engineering and quality manager, and reviewed by the managing director.

6 AUDITS

The engineering and quality manager will carry out regular audits on items in the quarantine area to ensure conformance to this procedure and make a report to the managing director.

COMPANY B

QUALITY PROCEDURE QPB 019

INTERNAL AND SUBCONTRACTOR QUALITY AUDIT

Compiled by...

Checked by...

Approved by...

Date...

CHANGE HISTORY PAGE

Document status	Date issued	Number of pages	Changed pages	Change/ defect no.
XX	XX	XX	XX	XX

CONTENTS LIST

1 INTRODUCTION

1.1 PURPOSE

The purpose of this procedure is to provide guidance on the planning and implementation of internal and subcontractor audits.

1.2 SCOPE

Quality audits will be conducted on all Company B's and subcontractor's procedures.

1.3 TERMS AND ABBREVIATIONS

None.

2 APPLICABLE DOCUMENTS AND REFERENCES

QMS 001—Quality Manual
QPS 003—Design and Review

3 GENERAL AUDIT REQUIREMENTS

Audits will be carried out as planned by the engineering and quality manager.

3.1 OBJECTIVES OF THE AUDIT

- To establish that there is adequate control over the design process.
- To establish that standards are being used for the documentation and production of the product.
- To seek evidence that the standards are being periodically reviewed.
- To establish that there are adequate controls over test.
- To establish that configuration control is applied.
- To establish that there is control of bought-in software.
- To verify that the Company B and subcontractor procedures are being applied.

Internal QUALITY AUDIT		
Department Audited:	Date of Audit:	Report No.: QA......................
Basis of Audit:		
Result/Non-conformance: Signed .... (Manager)		

....
(Date)   (Auditor)

....
(Date)   (Managing Director)

....
(Date)   (E. & Q. Manager)

Next audit due...

Fig. 1. Internal quality audit.

Internal QUALITY AUDIT Corrective Action		
Department Audited:	Date of Audit:	Report No.: QA........................

Corrective action:

Scheduled date of completion: Signed: ....
                                                                              (Manager)

Action pending: ....
                                          (Date)   (Auditor)

Action: Followed up: ....
                                          (Date)   (Managing Director)

Action: Complete: ....
                                          (Date)   (E. & Q. Manager)

Fig. 2. Internal quality audit, corrective action.

321

3.2 IMPLEMENTING THE INTERNAL AUDIT

The engineering and quality manager will maintain an internal audit file, which will contain:

—planned future audits (monthly)
—audit records
—outcome of remedial actions

The monthly audits will be planned in advance by the engineering and quality manager, who will maintain the appropriate plans in the audit file. These will address the implementation of the quality manual and procedures, and will review their application on a random sample basis. The audit plan will detail the features of the system that are to be audited.

The audit will be carried out by the engineering and quality manager and/or his representative (called the auditor). The audit will seek objective evidence that the specific procedures and standards are being implemented.

Formal audit records will be kept and will include:

—Reports of deficiencies (Fig. 1) with appropriate signatures.
—Target dates for remedial actions (Fig. 2) with appropriate signatures.

3.3 IMPLEMENTING THE SUBCONTRACTOR AUDIT

In the case of subcontractors, audits will include the vendor's quality organisation, documentation standards and quality procedures. The prime purpose will be to establish that he has a satisfactory quality system and, to that end, the existence of a quality manual will be compulsory. Evidence will be sought that manuals and procedures are being applied, and that they are not simply for show.

The first task should be to review the documents vertically to establish that requirements and functional specifications are fully reflected down to the module level of hardware and software.

Design module specifications should be audited for conformance to design standards, layout, requirements cross-referencing and functional performance. Coded software should be checked against any programming standards and cross-referred to the module design.

In most cases this will be conducted on a sample basis, in which case it is necessary to:

(a) Establish how long is to be spent on the activity so that the number of procedures and modules to be audited can be decided.

(b) Weight the time in favour of the requirements and functional specifications rather than skimping on this to permit more modules to be reviewed.

(c) Choose the sample of modules and procedures having regard to the critical features of the product (e.g. safety critical aspects).

(d) Include a sample of change and concession documents.

(e) Include a sample of documented design reviews.

The sequence should be:

(1) Plan
(2) Establish schedule of activities with subcontractor
(3) Prepare checklists
(4) Audit
(5) Initiate remedial action
(6) Prepare audit report

An important feature is that each and every deficiency should be recorded in writing and agreed, at the time, by everyone involved.

3.4 THE AUDIT REPORT

Where the audit is spread over a long time then interim reports should be prepared. At the end a full report should consist of:

(a) persons involved and their roles;
(b) any checklists used;
(c) written reports on each audited item;
(d) list of remedial actions;
(e) actions taken and modifications which resulted from them;
(f) an overview and recommendations.

3.5 SAMPLE CHECKLISTS

Checklists should be built up from past audits so that auditing is weighted towards those areas that traditionally cause problems.

QPS 003 and some other procedures contain some sample review checklists.

COMPANY C
QUALITY PROCEDURE QPC 001
CONTRACT REVIEW

Compiled by...

Checked by...

Approved by...

Date...

CHANGE HISTORY PAGE

Document status	Date issued	Number of pages	Changed pages	Change/ defect no.
XX	XX	XX	XX	XX

CONTENTS LIST

1 INTRODUCTION

1.1 PURPOSE OF DOCUMENT

The contract review is intended to ensure that all work performed for a customer complies with the terms of the contract between Company C and the customer.

1.2 SCOPE

Once a contract has been signed, Company C has made a commitment to produce a system, product or service for the client. This commitment applies to all contracts entered into by Company C.

1.3 TERMS AND ABBREVIATIONS

None.

2 APPLICABLE DOCUMENTS AND REFERENCES

BS 5750—(Part 1) British Standards, Quality Systems
QPC 004—Software Development
QPC 005—Software Design Standards
QPC 006—Software Review
QPC 009—Purchasing and Suppliers
QPC 015—Test
QPC 021—Servicing
QMC 001—Quality Manual

3 CONTRACT REVIEW

3.1 WHEN TO REVIEW

Contract reviews will be conducted on a regular monthly basis with every contract being reviewed at least once. For continuous maintenance and support and licence contracts, a review shall take place twice a year.

3.2 POINTS TO BE INCLUDED

It needs to be established that the following points have been taken into account and that Company C is complying with the requirements:

—the contract conditions
—standards and procedures called for by the contract
—acceptance criteria
—cost and time constraints
—risks and penalties
—resources needed

3.3 CONDUCT OF THE REVIEW

The contract review shall be conducted by a Company C director and the customer, and the main reference documents will be the contract and any other applicable documents.

The director will obtain the next available contract review number from the contract review list folder. A review form is then raised and should be completed as a form of minutes from the review.

On completion of the review the form will be filed in the review folder. If any corrective actions are required then the actions must be minuted and a future review date established to clear the outstanding actions.

3.4 CHECKLIST

—Have all the elements of the contract between Company C and the customer been completed?
—Does the contract cover ongoing support and/or maintenance?
—If so, does it meet the customer's needs?
—If not, does the customer require ongoing support or maintenance?
—Can Company C meet its obligations?
—Have any additional requirements been agreed to since the contract was signed?
—Have any standards called for been properly evaluated against Company C's capabilities?
—Do the conditions correctly apply to the product or service being supplied?
—Does the contract call for any specific suppliers or subcontractors?
—Is there a definitive list of deliverables (hardware and software)?

COMPANY C
QUALITY PROCEDURE QPC 002
CONFIGURATION AND CHANGE CONTROL

Compiled by...

Checked by...

Approved by...

Date...

CHANGE HISTORY PAGE

Document status	Date issued	Number of pages	Changed pages	Change/ defect no.
XX	XX	XX	XX	XX

CONTENTS LIST

1 INTRODUCTION

1.1 PURPOSE OF DOCUMENT

This procedure is intended to provide the control procedures for all changes that relate to documents and/or hardware and/or application software produced and/or provided by Company C. This procedure will conform to the BS 5750 (Part 1) British Standard for Quality Systems.

1.2 SCOPE

Company C has items which fall into three main areas:

(a) documentation
(b) hardware
(c) software

This procedure covers all of the change control procedures that must be applied to the above.

As it is the control of change that is being defined within this procedure, the source of the documentation, hardware or software can be from internal or external sources (i.e. the origination of the item need not be Company C).

For ease of use this procedure is structured in such a way that it follows the life-cycle of a normal project and it should be remembered that exceptions will exist where this life-cycle will not apply and some phases may be omitted. Further details of a project life-cycle are detailed in QMC 001 and QPC 004—Software Development.

Software excludes data as this is not subject to any form of change control. Only the software that manipulates the data can be effectively controlled.

1.3 TERMS AND ABBREVIATIONS

Documentation: any documentation produced for a project, be it internally produced or controlled by external owners.

Hardware: any physical computer equipment.

Software: any software that is used to provide a working system; this includes the operating system, package software and bespoke software but excludes data.

2 APPLICABLE DOCUMENTS AND REFERENCES

BS 5750 (Part 1) British Standards, Quality Systems
QMC 001—Quality Manual
QPC 008—Document Format and Standard
QPC 004—Software Development
QPC 005—Software Design Standards
QPC 015—Test

3 PROJECT LIFE-CYCLE

3.1 INITIAL CONTACT

The initial contact phase is not subject to change control procedures and is only present here for completeness of this document.

During the initial contact with a new client it is necessary to lay the foundations for the quality control of any product or service sold to that client.

During the initial contact phase a master customer record form is completed which summarises the client company and lists the contacts available in that company.

The initial contact phase, if successful, will result in the opening of a customer file.

The customer file will contain copies of *all* communications to the client, be it in the form of a letter, memorandum, proposal, specification or any other media. It is important that notes of all verbal communication are also kept in this file.

It should be remembered that the initial contact may have been initiated by the client.

3.2 IDENTIFICATION PHASE

The identification phase of a project is used to determine the project brief. This then determines the scope in which Company C can operate.

During the identification phase the system file is opened and all information relating to the requirements of the client are to be filed in the system file.

The only correspondence during the identification phase should be the production of a short report that identifies the scope and nature of the problem that requires resolution. It may be that the correspondence is in the form of a letter.

All such correspondence and reports relating to the proposed project are filed in the system file, and a copy of any letters sent to the client are also filed in the customer file.

If the correspondence is in the form of a report then it is subject to change control procedures defined later in this document.

3.3 PROPOSAL PHASE

The proposal phase is defined within this document as a separate phase to illustrate the importance of controlling the information contained within the proposal as it will form part of the contract between Company C and the client. The actual production of the proposal could be at the identification phase or the feasibility phase and is dependent upon the complexity of the solution.

The proposal is a very important document/report as it outlines the solution to an identified problem and the approximate costs involved in solving that problem. The accuracy of the costs is determined by the point at which the proposal is produced, i.e. the accuracy of the proposal is directly related to the effort used to investigate the problem.

It is very important, therefore, that disclaimers and assumptions are *always* included in any proposal produced by Company C.

3.4 FEASIBILITY PHASE

The feasibility phase results in the production of a feasibility report that details the problem, a single (or multiple) solution(s) and a recommendation, the recommendation being the preferred solution.

The feasibility report may be produced before or after the proposal and is always subject to change control procedures.

The feasibility report can only be released after a review by a senior member of the technical support staff. The intention of the review is to determine the viability of the recommendation, and to ensure that the report contains all the required elements.

After initial approval is granted then the proposal is given a 'controlled' document status and is subject to change control procedures.

3.5 ANALYSIS AND DESIGN PHASE

The analysis and design phase results in the production of a system description that details the solution to the problem defined by the feasibility report in more detail. The system description should be of sufficient detail to allow the construction phase to begin without the need for further major analysis.

The system description is always subject to change control procedures.

The system description can only be released after a review by a senior member of the technical support staff. The intention of the review is to ensure that the report contains all the required elements.

After initial approval is granted then the proposal is given a 'controlled' document status and is subject to change control procedures.

3.6 CONSTRUCTION PHASE

The construction phase involves the development of the system in line with the previously produced system description. Any documentary material produced at this stage will relate either to program specifications or will relate to changes required to the system description.

As the system description is subject to 'controlled' change control, refer to Section 4 on controlled change control procedures for more details.

During the construction phase the developer will be testing the functions produced and these tests are uncontrolled in that they are not checked by someone other than the author. However, before any function can be passed to the implementation phase it is subject to system testing procedures as defined in QPC 015—Test.

Once accepted for the implementation and acceptance phase all functions are subject to 'controlled' change control procedures.

3.7 TESTING PHASE

The testing phase involves the software undergoing a formalised testing procedure, after which, if successful, the functions then fall under 'controlled' change control procedures.

3.8 IMPLEMENTATION AND ACCEPTANCE PHASE

The implementation and acceptance phase involves the integration testing of the system and also user acceptance testing. As a large project will always be divided into sections then some of the implementation and acceptance phase may be completed before the end of the construction phase.

All functions are already under 'controlled' change control procedure at this stage.

During the implementation phase the user manual is produced and this must be approved by a senior member of the technical support staff. Once approved it is subject to 'controlled' change control procedures.

4 CHANGE CONTROL PROCEDURES

4.1 UNCONTROLLED DOCUMENTS AND FUNCTIONS

All uncontrolled documents and/or functions are not subject to any form of control and should only conform to the procedures and standards laid out in the quality procedures.

Because a function or document is not subject to change control procedures, it does not mean that it will not from time to time be checked for conformance to standards and other quality procedures. It simply means that such functions will not be formally reviewed.

All documents and/or functions, before any work is started upon them, must be allocated to a client. All clients are allocated a three-character mnemonic and this is recorded in the company identifier log.

Once allocated to a client the document and/or function must be allocated to a system. All systems within a client are allocated a unique two-character mnemonic and this is recorded in the system identifier log.

For documents a unique numeric identifier is allocated within the client and system identifiers, and recorded on the document identifier log and the document distribution log. That is, the issue number consists of two parts: the primary issue number, which is only incremented when the whole document must be reprinted, and the secondary issue number, which is only incremented when part of the document must be reprinted. If the primary issue number is incremented then the secondary issue number is always reset to zero.

A function name is allocated. The functions will then be issued a version number. The version number in relation to function is assigned to the whole system and not a part. The version number consists of three parts.

The primary version number is only incremented when the format and structure of the database over which the function operates is modified in any way. The secondary version number, or release level, is incremented when a new function is added to the system or a previous function is given additional functionality. The tertiary version number, or fix release level, is incremented when a correction is issued for an error within the system. If the primary version number is incremented then the secondary and tertiary version numbers are reset to zero. The information on software version numbers is recorded in the software version control and the software distribution log.

Examples of software version control:

(i) Initial review for a new function with acceptance of that function will result in the function being allocated a version number of 01.00.00.

(ii) An error is reported in the function via the error reporting system. This, on successful review of the modified function, results in a version number of 01.00.01.

(iii) A further error is reported in the function via the error reporting system. This, on successful review of the modification, results in a version number of 01.00.02.

(iv) A system amendment request for an enhancement to the function is approved and successfully reviewed, resulting in a version number of 01.01.02.

(v) A system amendment request for an enhancement which requires a change to the database is approved and successfully reviewed, resulting in a version number of 02.00.00.

4.2 CONTROLLED DOCUMENTS AND FUNCTIONS

All controlled documents and functions are subject to a formal review process before they can be set to controlled status and/or be allocated new issue or version numbers. The review process will be conducted by a senior member of the technical support staff and must be backed by the completion of a formalised review document.

Once a review has been completed the result can either be approval or rejection. If rejected, the document or function is returned to the originator for corrective measures and re-submitted at a later date. If accepted, the document is allocated a controlled status ('C') and can then be issued or re-issued as per the details contained in the document distribution list and/or the software distribution list.

4.3 FROZEN DOCUMENTS AND FUNCTIONS

Frozen documents and functions are controlled in the same manner as controlled documents and functions except that, as part of the review procedure, the client who has the rights to the document and/or function must also be involved in the review process. The intention of the frozen status is to prevent Company C from modifying a document or function without the knowledge of the client. This does not apply to standard software packages that are controlled by Company C.

An example of a frozen function would be if a client requested a specific modification to a package or specific bespoke function. Once reviewed and accepted it would be frozen and could not then be altered by Company C without a review with the client.

4.4 CONTROLLED HARDWARE

All hardware, by definition, is always controlled. The initial hardware is detailed in the proposal and should always be reviewed to ensure that it meets the requirements of the customer.

The original and all subsequent orders for hardware provide an audit trail together with the hardware configuration listings obtained from the computer system itself. The only mandatory number that must be obtained from the client is the hardware system serial number for all computer systems and this should be stored in the customer file. The hardware system serial number is used to provide a licence for the use of package software via a licence code which the user must enter into the computer system.

Expiry of the licence code or upgrading the computer system without the consent of Company C would therefore result in removal of the facility to run the packaged software.

5 APPENDICES

Appendices should be created to give examples of the documents mentioned in the text and listed on the contents page. These will be specific to your company.

COMPANY C
QUALITY PROCEDURE QPC 004
SOFTWARE DEVELOPMENT

Compiled by...

Checked by...

Approved by..

Date..

CHANGE HISTORY PAGE

Document status	Date issued	Number of pages	Changed pages	Change/ defect no.
XX	XX	XX	XX	XX

CONTENTS LIST

6 ANALYSIS AND DESIGN PHASE
   6.1 Objectives
   6.2 Activities
   6.3 Documentation
   6.4 Checkpoints
   6.5 Change Procedure

7 CONSTRUCTION PHASE
   7.1 Objectives
   7.2 Activities
   7.3 Documentation
   7.4 Checkpoints
   7.5 Change Procedure

8 TESTING PHASE
   8.1 Objectives
   8.2 Activities
   8.3 Documentation
   8.4 Checkpoints
   8.5 Change Procedure

9 IMPLEMENTATION AND ACCEPTANCE PHASE
   9.1 Objectives
   9.2 Activities
   9.3 Documentation
   9.4 Checkpoints
   9.5 Change Procedure

10 APPENDICES (examples to be created)
    Customer contact form
    Time sheet
    Customer file
    Identification report/proposal
    System file
    Activity control sheet
    Activity progress sheet
    Feasibility report
    Project progress report
    Data flow diagram
    Data model diagram
    Data dictionary relation sheet
    Data dictionary definition sheet

Screen layout
Screen reference sheet
Printer layout
Printer reference sheet
Function notes
System amendment request
Error report
System description
Technical file
User guide
Program structure diagram
Program structure action list
Function map

1 INTRODUCTION

1.1 PURPOSE OF DOCUMENT

The purpose of this procedure is to provide guidance on the development methods and controls to be used for all projects undertaken by Company C.

1.2 SCOPE

The scope is internal to Company C. If a customer wishes to impose different development procedures and practices then this must be reviewed with the technical director of Company C prior to any agreement to undertake such work.

1.3 TERMS AND ABBREVIATIONS

DFDs: data flow diagrams

2 APPLICABLE DOCUMENTS AND REFERENCES

BS 5750 (Part 1) British Standards, Quality Systems
QMC 001—Quality Manual
QPC 006—Software Review
QPC 005—Software Design Standards
QPC 002—Configuration and Change Control
QPC 015—Test
QPC 017—Non-conformance, Corrective Action and Records
Data Protection Act 1988

3 INITIAL CONTACT

3.1 OBJECTIVES

The initial contact with the client/prospect should serve to introduce Company C as an independent software consultancy and supplier—in particular the range of services and products that the company can provide.

At this stage the main client/prospect 'sponsor' should be identified. This sponsor will be the key figure in establishing a successful implementation of a project into the client site. The sponsor should be of sufficient authority to

QPC 004
Issue 1

ensure that key decisions are made without excess delay and therefore ideally have a board of directors position. In some organisations (e.g. local authorities, government, etc.) this will not be possible due to the tendering process and the best available sponsor should be obtained.

The Company C client support consultant for the client/prospect should be assigned at this stage. It will be the responsibility of the client support consultant to ensure that the applicable quality procedures and control mechanisms are complied with.

3.2 ACTIVITIES

The following activities should be initiated during the initial contact phase:

   (i) Allocate the client support consultant.
  (ii) Identify client/prospect 'sponsor'.
 (iii) Open customer file; if one does not already exist, complete the customer contact form.
 (iv) Minutes to be produced for all meetings and a copy maintained in the customer file.
  (v) Allocate a client reference, if one does not already exist. The client reference is a three-character code by which the customer will be known to Company C. The reference is made up by taking the first letter of each of the client's name up to a maximum of three. If the client's name does not consist of three parts then the next non-vowel (i.e. not a, e, i, o, u) is used, e.g.

      Bloggins Systems—BSS
      Harris Group—HGR

 (vi) Add time spent with client/prospect to time-sheets.

It is very important that time-sheets are completed at this stage for any work undertaken by the client support consultant and any other member of staff outside of the sales and marketing and administration departments. This then allows the cost of a sales activity to be established and therefore the total profitability of any project can be monitored.

3.3 DOCUMENTATION

Customer file to be opened and client contact form to be completed. The contents of the customer file should comply with that detailed in the appendix.

3.4 CHECKPOINTS

The following checkpoints should be considered during the initial contact phase:

(i) Is the task required suitable for Company C? That is to say, is the potential return from the work worthwhile and can Company C provide the service and/or product and maintain the level of quality required? (Refer to all quality procedures.)

(ii) Does the client/prospect wish to continue with Company C, and does the client/prospect have a requirement for a service or a product?

3.5 CHANGE PROCEDURE

Not applicable.

4 IDENTIFICATION PHASE

4.1 OBJECTIVES

This phase is intended to identify, in outline, the work required and to detail our terms of reference. At this stage this will probably be only a few days (or even hours) work. The work can be carried out by the sales person or the client support consultant, or both.

4.2 ACTIVITIES

The following activities should be initiated during the identification phase:

(i) Initial interviews with user management.

(ii) Produce outline costs of software and/or hardware requirements for user.

(iii) Cost the feasibility study phase (if applicable).

(iv) Initiate a system file and allocate a system reference. The system reference is a two-character code made up of the first character of the names of the system and also applying the same rules as that of the client reference, e.g.

- product system—PS
- general accounting system—CG
- commercial system—CS

It is allowed to have duplicate system codes provided it does not occur within the same client. The system file will contain all project-related correspondence. However, a copy of any letters sent to the client will also be filed in the customer file.

(v) Presentation to senior users on identification reports.

(vi) Signing of contract.

(vii) All minutes to be produced and a copy maintained in both system and customer files.

(viii) Produce separate time-sheets for client.

(ix) Note any specific performance and response-time requirements.

4.3 DOCUMENTATION

At this stage an identification report may be needed; this should outline the possible solutions to be investigated further in the next phase. The identification report may also be called the proposal. The contents of the identification report or proposal should comply with that detailed in the appendix.

Ensure that all correspondence is filed correctly; all project-related correspondence in the system file and copies of all minutes and letters in the customer file. The contents of the system file should comply with that detailed in the appendix.

4.4 CHECKPOINTS

The client/prospect must agree to either the costing of the feasibility study and/or the costing of the product(s) detailed within the proposal. The client/prospect must then sign a contract with Company C before the next phase can commence. The client/prospect may at this phase decide not to proceed.

4.5 CHANGE PROCEDURE

Once a proposal has been produced by the client support consultant and/or the sales person then it *must* be reviewed and approved prior to distribution to the client/prospect (refer to QPC 006—Software Review).

Once approved the proposal is then subject to change control procedures and cannot be modified without the need for a further review (refer to QPC 002—Configuration and Change Control).

It is the responsibility of the client support consultant to ensure that the review and configuration procedures are carried out.

344

5 FEASIBILITY STUDY PHASE

5.1 OBJECTIVES

The feasibility study may sometimes be referred to as the 'package fit study'. In such cases the objectives are the same but the resultant document will recommend changes to an existing package/product to meet the requirements of the client/prospect.

The prime objective is to weigh up the alternative solutions, to recommend one, and for the client to agree to that proposal.

At this point we must also ensure that commitment is demonstrated on both sides; the client must be made aware of all future commitments, such as reading documentation, providing access to key personnel, etc. Commitment must also be demonstrated by Company C to show that we are a truly professional organisation and that the solution proposed is achievable.

5.2 ACTIVITIES

The following activities should be initiated during the feasibility phase:

(i) Further analysis of the work, leading to initial outline of data flow diagrams and the data model. Refer to appropriate appendix for detailed description of data flow diagrams and data model requirements. The analysis can be carried out by the client support consultant or a systems analyst, dependent upon the nature of the task.

(ii) Detail hardware required for the solution proposed, if appropriate.

(iii) Initial consideration of recovery and fall-back requirements, if appropriate.

(iv) Produce detailed costing of the analysis and design phase, and outline of costings for the remainder of the project. Also establish time-scales and produce an outline project schedule.

(v) Present findings of the study to senior user management for their approval and subsequent signing.

(vi) Define the main activities required to complete the whole project and complete an activity control sheet (see appendix) for each of the activities. These activities will become the milestones of the project and will be used for monitoring. Some of the individual tasks within an activity may not be known in detail at this phase and will be completed as more information is available.

345

(vii) Complete an activity progress sheet (see appendix) on a weekly basis for the duration of the feasibility study.

(viii) Decide on the use of any specific tools.

(ix) Note any specific performance and response-time requirements.

5.3 DOCUMENTATION

The feasibility report or package fit report must be produced, along with the costings and the schedule of activities, as defined by the activity control sheets. The contents of the feasibility report or package fit report should comply with that detailed in the appendix.

5.4 CHECKPOINTS

The client should sign the feasibility report and agree to the proposals before continuing to the next stage.

5.5 CHANGE PROCEDURE

Once a feasibility report or package fit study has been produced by the client support consultant then it *must* be reviewed and approved prior to distribution to the client (refer to QPC 006—Software Review).

Once approved the report/study is then subject to change control procedures and cannot be modified without the need for a further review (refer to QPC 002—Configuration and Change Control).

If changes to the requirements are introduced at this stage by the client, the client support consultant must decide if these changes can be included directly at this stage, or if a new identification phase must be started. A new identification phase should only be necessary if the change is major, but the client support consultant must make the client aware of the consequences of the changes, including changes to the costings and schedule.

6 ANALYSIS AND DESIGN PHASE

6.1 OBJECTIVES

The prime objective of this phase is to define, in detail, the work to be done in the construction phase. At the end of this phase, the system description must be completed by Company C and agreed by the client.

346

It is frequently advisable to break this phase into two parts. The first will be concerned with the overall approach to the system, while the second will fill in all the detail. The main point is to ensure that decisions such as hard-coding of data and the break-down of the system into separate sub-systems are taken as early as possible, thus avoiding major changes to the design later during this phase.

6.2 ACTIVITIES

The following activities should be initiated during the analysis and design phase:

(i)   A project leader is allocated and, if necessary, a project team is provided for the completion of the analysis and design phase.

(ii) Define the frequency, venue and attendees of project progress meetings at the start of the phase. Also define the distribution list for minutes and other papers produced during this phase. All meetings are to be minuted and distributed in accordance with the distribution list.

(iii) Complete project progress reports (see appendix) at user-defined intervals, although the maximum interval must not exceed one month.

(iv) Detail the data flow diagram (see appendix) and the data model diagram (see appendix).

(v)   Create data dictionary using the forms provided (see appendix).

(vi) Design all screens and report layouts using the forms provided (see appendix).

(vii) Design all other functions required, including any manual functions, such as checking procedures. Use the forms provided, if appropriate (see appendix).

(viii) For complex arrangements of nested functions provide a map of the functions available. Use function map form (see appendix).

(ix) Produce detailed costing and schedules for the remainder of the project—the construction and implementation phases.

(x)   As an internal exercise, check the hardware sizing for the client. This need only be released to the client on request, or if we think there are potential problems. Refer to QPC 006—Software Review for further details.

(xi) Design menus and user security considerations.

6.3 DOCUMENTATION

The full system description will be produced, along with full costings and schedules, for the remainder of the project. Refer to the appendix for the contents of the system description.

The technical file (for internal use), containing the technical manual, will also be started; this will contain all information required to create the system that only the development team need consider. Refer to the appendix for the contents of the technical file.

6.4 CHECKPOINT

The client must sign the system description and agree the further costings before continuing.

6.5 CHANGE PROCEDURE

Once a system description has been produced by the systems analyst or project leader then it *must* be reviewed and approved prior to distribution to the client (refer to QPC 006—Software Review).

Once approved the report is then subject to change control procedures and cannot be modified without the need for a further review (refer to QPC 002—Configuration and Change Control).

Again, if changes are made to the requirements during this phase, it is the responsibility of the project leader to decide if the changes can be included at this stage, or if a new identification phase and/or feasibility study is required. In any event, if changes are required all costs and schedules must be updated and the client must authorise the changes.

These change requests should be documented using the system amendment request forms (see appendix). The change cannot refer to the system description itself if it is not complete as it refers to the feasibility study or package fit report. However, if the system description is in a controlled state (i.e. it has been reviewed and approved) then the system amendment request refers to the system description. All such changes must be reviewed.

7 CONSTRUCTION PHASE

7.1 OBJECTIVES

To create the system as described in the system description, to the point where it is ready for hand-over to the user.

348

7.2 ACTIVITIES

The following activities should be initiated during the construction phase:

(i) Create all files and programs. Use the program structure diagram and program structure action list, if appropriate.
(ii) Create 'help' text.
(iii) Create user guide.
(iv) Test functions within Company C.

7.3 DOCUMENTATION

The system description will have been changed during this phase (with accompanying changes in the costs and schedules) if the user requirements have changed. Any change must be accompanied by an approved system amendment request and the associated budget costs are increased accordingly. The system amendment request *must* be signed by an authorised signatory for the client for it to be valid.

A user guide, including the operational instructions, will be produced if the client has requested them.

Note that the technical file, containing the technical manual, initiated in the previous phase may well be expanded for internal information purposes, even if the client does not require it.

All functions will have a working program file opened during this phase. The contents of the program file will be the program test plan and any results from program testing.

All changes introduced must be accompanied by a system amendment request form, duly authorised.

7.4 CHECKPOINTS

The project manager must sign the system off as being completed and fully system-tested. This will require the setting up of a test-pack. This test-pack should, if possible, be made up of data provided by the users themselves, as well as any test data created internally to test all events.

7.5 CHANGE PROCEDURE

If changes are introduced by the client during construction, they must be re-costed by the project leader and agreed by the client, using a system amendment request form. Only then can they be included in the project. Refer to QPC 002—Configuration and Change Control for more details.

8 TESTING PHASE

8.1 OBJECTIVES

The objective of the testing phase is to system-test the product produced by the construction phase and to ensure that it meets the requirements of the client.

8.2 ACTIVITIES

The following activities should be initiated during the testing phase:

(i) System-test in accordance with QPC 015—Test.

(ii) Produce error reports (see appendix) for areas within the system not complying to the requirements of the client. The error reports will use a numbering sequence specific to that client and will therefore be prefixed with the client reference code. Such error reports are passed back to the developer for correction.

(iii) Once the system test has been completed and the system complies with the requirements of the client it is reviewed and, if successful, the final implementation stage can commence.

8.3 DOCUMENTATION

The only documentation created at this stage of the project should be error reports to correct faults within the system.

However, if an error is the result of a mistake or logic fault within the system description then a system amendment request must be raised and reviewed to instigate a change to a controlled document (i.e. the system description). Normal errors in the functions that do not comply with the system description need only an error report to correct whilst the system is in an uncontrolled state (see Section 8.5).

8.4 CHECKPOINTS

The system must pass the test review to be allowed to progress to the next phase.

8.5 CHANGE PROCEDURE

Whilst the system functions are not under change control they can be amended to remove errors via the completion of an error report. However, once the system has been successfully reviewed then it transfers to controlled status and cannot be changed without the completion and approval of a system amendment request.

If changes are introduced by the client during testing, they must be re-costed by the project leader and agreed by the client, using a system amendment request form. Only then can they be included in the project.

Any changes introduced due to logic errors in the system description must be reviewed as to the cause of the original error; if the error is a result of the client providing incorrect information then they will result in a re-costing of the project.

9 IMPLEMENTATION AND ACCEPTANCE PHASE

9.1 OBJECTIVES

To hand the finished system over to the users, have them test it to their satisfaction and sign off the project as complete.

9.2 ACTIVITIES

The following activities should be initiated during the testing phase:

- (i) Hand over a complete copy of the system to the user.
- (ii) Assist, as required by the client, with the implementation of the live system.
- (iii) Assist with, and oversee, all user testing and parallel running of the system.
- (iv) Archive the system for future reference and support, if not written on the client's own machine.
- (v) Finally, when the project is signed off, an end review should be held, with a written report, to conclude the project.

9.3 DOCUMENTATION

The only documentation created at this stage of the project should be the written version of the end review.

9.4 CHECKPOINTS

The client 'sponsor' must sign the project off as being completed and installed to their satisfaction. At this point the project is considered to be finished and any further work will initiate a new project.

9.5 CHANGE PROCEDURE

Changes may not be introduced at this stage as the system has already been agreed by the client. However, any errors on live running will have to be corrected. Any actual changes to the requirements at this stage must be considered as extra to the project, so effectively become their own 'mini-project'.

10 APPENDICES

Appendices should be created to give examples of the documents mentioned in the text and listed on the contents page. These will be specific to your company.

COMPANY C
QUALITY PROCEDURE QPC 005
SOFTWARE DESIGN STANDARDS

Compiled by...

Checked by...

Approved by...

Date...

CHANGE HISTORY PAGE

Document status	Date issued	Number of pages	Changed pages	Change/ defect no.
XX	XX	XX	XX	XX

CONTENTS LIST

1 INTRODUCTION

1.1 PURPOSE OF DOCUMENT

This procedure is intended to provide the standards by which all programs are generated within Company C.

1.2 SCOPE

Company C has a number of available programming languages:

  (a) RPG III
  (b) Synon/2
  (c) COBOL
  (d) CL

This procedure covers the standards to be used in writing or programming in any of the nominated languages.

The programming standards referred to by this procedure only refer to work that is done in-house and if a client wishes to use his/her own programming standards then they are substituted for this procedure.

1.3 TERMS AND ABBREVIATIONS

RPG III: report program generator, version III.

Synon/2: a computer-aided systems engineering (CASE) tool developed by Synon Limited and now an IBM product.

CL: control language is that which is used to control the running of functions on the computer system.

In-house: any work that is performed by Company C at the main offices.

CASE: computer-aided system engineering.

2 APPLICABLE DOCUMENTS AND REFERENCES

BS 5750 (Part 1) British Standards, Quality Systems
QMC 001—Quality Manual
QPC 004—Software Development
QPC 002—Configuration and Change Control

QPC 008—Document Format and Standard
Synon/2 Reference Manual
IBM Guide to SAA

3 NAMING CONVENTION FOR OBJECTS

3.1 AS/400 AND S/38 NAMING CONVENTION

The following naming convention is used for objects' names:

SSABBOOQ

where

SS = system identifier (e.g. PS = product system; CS = commercial system).

> Note: the sub-system identifier will be the same as the system identifier allocated to the system description document and recorded in the document recording logs

A = type of function, selected from

> M = file maintenance program
> C = communications program
> F = data file
> V = validation program
> A = menu program
> E = enquiry program (including reports)

BB = unique mnemonic for the function or file within system

OO = type of object may be one- or two-character, selected from

> R = RPG program
> CB = COBOL program
> B = BASIC program
> P = PASCAL program
> F = FORTRAN program
> C = CL program
> CP = C program
> PF = physical file
> LF = logical file
> DF = display file
> SF = print file

Q = qualifier which is used if more than one function or files are required to be associated together, e.g. (a) an on-line CL program calls a batch CL program giving the names PSEALC and PSEALC1 respectively, and (b) a physical file (PSFCMPF) has two logical files (PSFCMLF1 and PSFCMLF2)

3.2 SYNON/2 NAMING CONVENTION

Synon/2 is a computer-aided system engineering (CASE) tool and as such has an in-built set of rules for the naming of all objects generated from such a facility. Refer to the Synon/2 Reference Manual for further details.

The naming convention used externally to Synon/2 is sufficiently different as to not cause a conflict in names if a Synon/2 generated system and a 'manually' generated system were required to co-exist in the same environment.

4 STRUCTURED PROGRAMMING

Structured programming is important for a number of reasons:

from	**S**—standards
giving	**T**—traceable code
	R—readable code
	U—understandable programs
	C—common coding techniques
through	**T**—top-down design
which is	**U**—upgradable
	R—reliable
	E—easy to maintain

With all languages available to Company C it is possible to use structured programming techniques.

Within structured programming there are three basic elements required to write any program:

(a) a sequence —any action that must be performed
(b) an iteration—any set of sequences that must be performed a number of times
(c) a selection —a set of sequences that are conditional

The symbols to be used in structural programming are as follows:

(a)   a sequence:

1

where 1 is a step, which may be a single instruction or a series of instructions.

(b)   an iteration:

1

2 *

where 2 is the iterated part of step 1.

(c)   a selection:

3

4 0	5 0

where 4 and 5 are the selections based upon the tests performed at 3.

The symbols used are defined as follows:

'*' indicates that this element is an iteration of the previous element in accordance with the limit specified between the two elements.

'0' indicates that this element is a selection of the previous element in accordance with the limit specification of that element.

Each diagram will have a number of actions defined and these actions are recorded in the action list. An example is shown opposite.
The associated action list would be:

0.1—start of function
1.1—open files and read first record
1.2—iteration until EOF reached
2.1—iterated part
3.1—selection of type A records only
4.1—selected part of type A
5.1—accumulate values and obtain description from codes provided on record
5.2—print record details

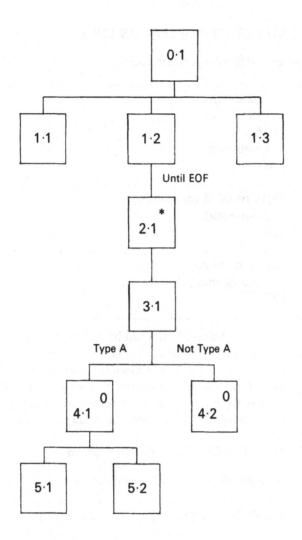

4.2—selected part of NOT type A

1.3—print 'End of Report' and close files

A number of forms are provided to aid in the production of the program structure diagram and associated action lists.

It is not the intention of programming standards or procedures to define in detail the techniques of structured programming as many reference documents and books exist for this purpose. However, all programming should be of a structured nature.

4.1 RPG III STRUCTURED OPERATION CODES

RPG III has available the operation codes:

....

```
IF (condition)
    [sequence]
ELSE
    [sequence]
END
```

....

```
DOW or DOU (condition)
    [sequence]
END
```

....

```
CAS (condition)
        [sequence]
END
```

....

Every program has a hierarchy and can be broken down into modules that can be easily understood and maintained. A module may be a sub-routine or, in some cases, a separate program. Design the modules to perform a single function or a number of highly related functions (cohesion). Minimise the inter-reactions between modules (coupling); this not only helps maintenance but enhances the chance of re-usability.

4.2 COBOL STRUCTURED OPERATION CODES

COBOL has available the structured operations codes:

....

```
PERFORM [sequence] UNTIL {condition}
```

....

```
PERFORM [sequence] WHILE {condition}
```

....

4.3 PROGRAM STRUCTURE

As a minimum the program should be divided into the following modules:

(a) initialisation
(b) main body
(c) termination

These sections are mandatory in all programs. However, as Synon/2 is already structured in this fashion the rule does not apply.

All programs should have a standard header section. The following example is for an RPG III program but the same applies to a COBOL program (only the format of comment lines being different):

```
*********************************************************************
* (c) 1991—Company C
*********************************************************************
* Client...................... :   Company D
* System..................... :   Product System
* Program i.d. .............. :   PS010                    Version 01.00.02
* Program.................... :   Cross reference report by product status
* Specified by.............. :   A. N. Other              Date: 10/03/89
* Written by ................ :   B. N. Other              Date: 14/03/89
* -----------------------------------------------------------------
* Description............... :   The print program will summarise the products by
                                 status, giving the total products that are within each
                                 category.
* -----------------------------------------------------------------
* Sub-routines............. :   STDHDR—standard report header
* Calls....................... :   PRTALC—printer allocation and override
* Indicators ................ :   01–24 Command keys
*                                    25 Help
*                                    26 Roll Up
*                                    27 Roll Down
*                                    28 Home
*                                    29 Print Key
*                                 30–39 Subfile processing
*                                 40–69 Record processing
*                                 90–99 File Errors
* -----------------------------------------------------------------
* Amendments:
*    No ...................... :   A001
*    Change ref. .......... :   CCSERRO1/124
*    Date ................... :   17/03/89
*    Amended by ........ :   A. N. Other
*    Reason................. :   The report was incorrectly sequenced and was using
*                                 the wrong logical view.
*    No. ..................... :   A002
*    Change ref. .......... :   12345
*    Date ................... :   17/04/89
*    Amended by ........ :   A. N. Other
*    Reason................. :   Customer request to change the layout of the report.
*********************************************************************
```

The header section should also contain details of all amendments to the program and all amended lines should hold a reference to the amendment number. This is needed for traceability of modifications to functions.

At least 35% of a program should consist of comments noting the processing that the program performs and the reason for the process.

5 PROGRAMMING CONVENTIONS

5.1 NAMING CONVENTIONS WITHIN RPG III

All field names within RPG III must be no greater than six characters. It is difficult, therefore, to have meaningful names. The following are guidelines to be used:

mmmmmm

where mmmmmm = a meaningful mnemonic for the field.

However, if an array is needed then the field name must be reduced to four characters.

Any internal work fields required by a function should be prefixed with a 'W'.

5.2 NAMING CONVENTIONS WITHIN COBOL

COBOL allows a large field name and all field names used from a file should be mapped to a meaningful name. It will not be possible to allow data names within files to exceed six characters so that RPG III programs can still be used if required.

All work fields used within a function must be prefixed with a 'Wnn', where 'nn' indicates the level within the working storage section of the COBOL program.

5.3 NAMING CONVENTIONS WITHIN SYNON/2

Synon/2 has an automatic naming convention for files and fields, and should not be altered in any way. However, these names, which conform to RPG III standards, are mapped to an internal name that is then used within the Synon/2 model for the generation of functions.

As Synon/2 defines screen formats automatically and uses as default the mapped name of the item as a heading, then whenever possible the data dictionary should be as meaningful as possible and the use of prefixes and suffixes discouraged.

However, at the stage of systems design, if the data dictionary is complex then it may be necessary to use a prefix. In such a case the real name of the item should also be recorded and the default label within the Synon/2 model changed. (Note that such an exercise is very time consuming prior to the generation of any function.)

Any work fields that require generation during the construction of the Synon/2 system should be prefixed with a # (hash), and be referenced whenever possible to fields already within the model.

5.4 SCREEN LAYOUTS

All screen layouts should conform to the appropriate system application architecture standard (e.g. as defined by IBM). Details of this standard can be found in the IBM Manual: A Guide to SAA.

5.5 REPORT LAYOUTS

All reports should have a header and also a trailer to every report. The header must contain the name of the function generating the report, the name of the report, the company name, and the date and time of production of the report. The trailer of the report should contain the line '***** End of Report *****'.

If the report has a number of user-defined selection criteria then the selection criteria must be printed once at the beginning of the report.

COMPANY C

QUALITY PROCEDURE QPC 006

SOFTWARE REVIEW

Compiled by...

Checked by..

Approved by..

Date...

CHANGE HISTORY PAGE

Document status	Date issued	Number of pages	Changed pages	Change/ defect no.
XX	XX	XX	XX	XX

CONTENTS LIST

7 IMPLEMENTATION REVIEW
   7.1 General
   7.2 When to Review
   7.3 Conduct

APPENDIX 1: REVIEW CONTROL LIST
APPENDIX 2: REVIEW FORM
APPENDIX 3: REVIEW CHECKLISTS

1 INTRODUCTION

1.1 PURPOSE OF DOCUMENT

This procedure provides the review mechanism for use by Company C. The word 'review' is defined as:

a viewing again: a looking back, retrospect: a reconsideration: a survey: a revision: a critical examination: a critique...

It is intended to be the medium whereby the hardware configuration, software configuration, design and development of a system shall be critically examined and approved. This will establish discrete points of achievement and provide confidence that the total system solution will meet the customer's requirements.

1.2 SCOPE

The document is to cover all review procedures for hardware configuration, software configuration, design, development, testing, implementation and contract. The design and development only relate to bespoke systems or bespoke amendments to existing systems.

All subcontracted software products are reviewed via the test review procedure as no direct control can be obtained over the design and development stages.

1.3 TERMS AND ABBREVIATIONS

Profile: a profile is a measured activity on a computer for a specific type of work load (e.g. word processing, printing, high speed data entry, etc.).

Software configuration: any group of computer programs that are required to make a computer function to the requirements of the customer.

Hardware configuration: any physical item of computer equipment.

2 APPLICABLE DOCUMENTS AND REFERENCES

BS 5750 (Part 1) British Standards, Quality Systems
QMC 001—Quality Manual
QPC 001—Contract Review

QPC 008—Documentation Format and Standard
QPC 004—Software Development
QPC 005—Software Design Standards
QPC 002—Configuration and Change Control
QPS 007—Subcontract Equipment and Software
QPC 015—Test
QPC 017—Non-conformance, Corrective Action and Records

3 HARDWARE AND SOFTWARE SYSTEM CONFIGURATION REVIEW

3.1 GENERAL

The purpose of the hardware and software system configuration review is to ensure that the customer's requirements are met in terms of:

(i) functionality of system
(ii) response times of system
(iii) growth of system

3.2 WHEN TO REVIEW

An activity flowchart, which highlights the points at which hardware and software system configuration reviews should take place, will be prepared by the technical director for each project.

In summary, a hardware and software system configuration review should occur prior to the release of any documentation or correspondence that details a recommendation by Company C to a customer that affects the customer hardware and/or software system requirements.

3.3 SIZING

As part of the hardware and software system configuration review a sizing exercise *must* be performed using the best tool/mechanism available.

Sizing is an imprecise science and by its very nature is approximate. The only practical approach is to benchmark a system, which is only feasible in existing customer situations. New situations and customers cannot have benchmarks performed without great expense, which is not justified in the great majority of cases. Other methods of sizing a system must therefore be used and to provide good sizing require:

(i) attention to detail
(ii) judgement
(iii) experience
(iv) understanding of the IBM sizing tools, if in an IBM environment
(v) understanding of the customer's requirements
(vi) effort
(vii) iteration and review
(viii) existing similar situations

Examples of IBM tools available for system sizing are listed below in order of preference of use:

(i) MDLSYS
(ii) QSIZE400
(iii) SSG400

3.3.1 MDLSYS

MDLSYS is part of the IBM AS/400 performance measurement tools set and is an analytic modelling/capacity planning tool.

3.3.2 QSIZE400

QSIZE400 is available on Dial-IBM and is based upon a table lookup with some analytic modelling.

QSIZE400 is intended for initial sizing and as such has been designed for ease of use.

QSIZE400 provides full documentation as part of the tool, including standard disclaimers, which can be used as an appendix to a proposal document.

3.3.3 SSG400

SSG400 is a PC-based tool that is simple to use and is intended for initial sizing.

3.4 CONDUCT

The hardware and software system configuration review should be conducted by a senior member of the technical support department, and both the salesperson and the customer support consultant responsible for the customer should also be present.

The relevant checklist(s) should be used to conduct the review, it being the responsibility of the senior member of the technical support department to ensure that the latest issue of the checklist(s) are used.

The senior member of the technical support department will obtain the next available review number from the review control list folder (an example of a review control list can be seen in Appendix 1). A review form (Appendix 2) is then obtained and should be completed together with minutes of the review detailing the comments on each of the points on the checklist(s).

On completion of the review, the review form and review minutes are filed in the review folder. The results on the review will then be recorded on the review control list. If any corrective action is required, the action depends upon the control status of the document being reviewed.

An uncontrolled document will be amended and be subject to a further review once the points noted on the review form have been addressed.

A controlled document will require the raising of a system amendment request form before it can be amended.

Upon a successful review the document detailing the hardware and software system configuration is raised to controlled status.

4 DESIGN REVIEW

4.1 GENERAL

All systems should be designed in accordance with the development procedures and take full advantage of structured data analysis techniques. It is the purpose of the design review to ensure that the proposed design conforms to the standards and will meet the customer's requirements.

A design review is internal to Company C and applies to every design project undertaken. A design review will take place on completion of the following stages (referred to in QPC 004—Software Development):

(i) feasibility study
(ii) functional specification
(iii) system description

The functional specification may sometimes be referred to as a 'package fit study', in that the requirements of a customer are reviewed and matched against a standard package and corresponding recommendations made.

4.2 WHEN TO REVIEW

A review will be conducted on completion of a feasibility study, functional specification or system description but prior to publication of the document.

Prior to its first review all design documents are always in an uncontrolled status. After a successful review they become controlled. Refer to QPC 002—Configuration and Change Control for further details.

4.3 CONDUCT

The design review should be conducted by a senior member of the technical support department and the originator of the document, and any other party with an interface to the document.

The main reference documents for conducting the review are:

(i)   QPC 004—Software Development
(ii) any customer documentation, requirements specification, etc.
(iii) design review checklist (see Appendix 3)

The senior member of the technical support department will obtain the next available review number from the review control list folder (an example of a review control list can be seen in Appendix 1). A review form (Appendix 2) is then obtained and should be completed together with minutes of the review detailing the comments on each of the points on the design review checklist.

On completion of the review, the review form and review minutes are filed in the review folder. The results on the review will then be recorded on the review control list. If any corrective action is required, the action depends upon the control status of the document being reviewed.

An uncontrolled document will be amended and be subject to a further review once the points noted on the review form have been addressed.

A controlled document will require the raising of a system amendment request form before it can be amended.

Upon a successful review the document detailing the hardware and software system configuration is raised to controlled status.

371

5 DEVELOPMENT REVIEW

5.1 GENERAL

The development review is intended to ensure that all development standards and procedures are being adhered to in the correct manner. The quality procedures QPC 004—Software Development and QPC 005—Software Design Standards are used within the development review process.

5.2 WHEN TO REVIEW

A development review will be conducted on a formal basis, with every bespoke system/project being reviewed at least once; larger systems will be reviewed quarterly. The definition of a larger system is any system that requires more than six months elapsed effort.

5.3 CONDUCT

The development review should be conducted by a senior member of the technical support department and the project leader assigned to that project.

The main reference documents for conducting the development review are:

  (i) QPC 004—Software Development
  (ii) QPC 005—Software Design Standards
  (iii) development review checklist (see Appendix 3)

The senior member of the technical support department will obtain the next available review number from the review control list folder (an example of a review control list can be seen in Appendix 1). A review form (Appendix 2) is then obtained and should be completed together with minutes of the review detailing the comments on each of the points on the development review checklist.

On completion of the review, the review form and review minutes are filed in the review folder. The results of the review will then be recorded on the review control list.

If any corrective actions are required then the actions are minuted and a future review date is established, when the project can be reviewed again for compliance.

6 TEST REVIEW

6.1 GENERAL

The test review is intended to ensure that all testing standards and procedures are being adhered to in the correct manner. The quality procedure QPC 015—Test is used within the test review process.

6.2 WHEN TO REVIEW

A test review will be conducted on a formal basis, with every bespoke system/project being reviewed at least once; larger systems will be reviewed quarterly. The definition of a larger system is any system that requires more than six months elapsed effort.

It is allowed to perform the test review at the same time as the development review and to combine the results of the reviews.

6.3 CONDUCT

The test review should be conducted by a senior member of the technical support department and the project leader assigned to that project.

The main reference documents for conducting the test review are:

(i) QPC 015—Test
(ii) test review checklist (see Appendix 3)

The senior member of the technical support department will obtain the next available review number from the review control list folder (an example of a review control list can be seen in Appendix 1). A review form (Appendix 2) is then obtained and should be completed together with minutes of the review detailing the comments on each of the points on the test review checklist.

On completion of the review, the review form and review minutes are filed in the review folder. The results of the review will then be recorded on the review control list.

If any corrective actions are required then the actions are minuted and a future date is established, when the project can be reviewed again for compliance.

7 IMPLEMENTATION REVIEW

7.1 GENERAL

The implementation review is intended to ensure that all implementation procedures are being adhered to in the correct manner. The quality procedure QPC 004—Software Development is used within the implementation review process.

7.2 WHEN TO REVIEW

An implementation review will be conducted on a formal basis, with every bespoke system/project being reviewed at least once; larger systems will be reviewed quarterly. The definition of a larger system is any system that requires more than six months elapsed effort.

It is allowed to perform the implementation review at the same time as the development review and the test review, and to combine the results of the reviews.

7.3 CONDUCT

The implementation review should be conducted by a senior member of the technical support department and the project leader assigned to that project.

The main reference documents for conducting the implementation review are:

(i) QPC 004—Software Development
(ii) implementation review checklist (see Appendix 3)

The senior member of the technical support department will obtain the next available review number from the review control list folder (an example of a review control list can be seen in Appendix 1). A review form (Appendix 2) is then obtained and should be completed together with minutes of the review detailing the comments on each of the points on the test review checklist.

On completion of the review, the review form and review minutes are filed in the review folder. The results of the review will then be recorded on the review control list.

If any corrective actions are required then the actions are minuted and a future review date is established, when the project can be reviewed again for compliance.

APPENDIX 1: REVIEW CONTROL LIST

Review Control List						
Review number	Client ref.	System ref.	Date assigned	Review date	Pass/ fail	Comments

Appendix 2 follows

APPENDIX 2: REVIEW FORM

Review
Review reference........................ Page: 1 of
Review title(s) ....

The above named document(s)/function(s)* are subject to review. The aim(s) of the review are to:
  (a) approve the document(s)/function(s) for transfer to controlled' change control procedures.
  *(b) agree that the design of system meets the client's requirements.
  (c) approve the document(s)/function(s) for re-issue.
  (d) agree the proposed changes as defined by the attached error log/system amendment request and any additional actions required.
  *(e) other; please specify:
 	
Please review the attached document(s)/function(s)* by / /
Originator:
  Signature............................ Date................................ / /
 	
  Name ....

Taking into account the aims of the review:
  (a) I have no objections to the document(s)/function(s), subject to comments overleaf.
  (b) I have objections to the document(s)/function(s) which are detailed below.
  (c) I have objections to the document(s)/function(s) and consider a formal meeting is necessary.
 	
Reviewer:
  Signature............................ Date................................ / /
 	
  Name ....

Comments: Note: *delete as necessary.

APPENDIX 3: REVIEW CHECKLISTS

DESIGN REVIEW CHECKLIST

(1) Does the design match the customer requirement?
(2) Does the design conform to the standards defined in QPC 004—
Software Development?
    (2.1) If not, does it conform to the requirements defined by the customer?
(3) Has the design been broken into activities and tasks within activities for the construction phase?
(4) Are the estimated time-scales for the construction phase realistic?
(5) Can Company C complete the construction phase in the time-scales?
(6) Is additional resource needed for the project?

DEVELOPMENT REVIEW CHECKLIST

(1) Are the functions being written to the standards defined within QPC 005—Software Design Standards?
(2) Is the technical file being kept up to date?
(3) Are function files being maintained for all functions?
(4) Do program test plans exist?

TEST REVIEW CHECKLIST

(1) Has a system test plan been prepared?
(2) Do the expected test results match the actual test results?

IMPLEMENTATION REVIEW CHECKLIST

(1) Has the user manual been completed?
(2) Is the customer happy with the quality of the software/product?
(3) Does the customer understand the error recording and change control mechanism for recording of errors and requesting a change?
(4) Has a hand-over mechanism been agreed with the customer?

COMPANY C

QUALITY PROCEDURE QPC 008

DOCUMENTATION FORMAT AND STANDARD

Compiled by...

Checked by...

Approved by...

Date...

CHANGE HISTORY PAGE

Document status	Date issued	Number of pages	Changed pages	Change/ defect no.
XX	XX	XX	XX	XX

CONTENTS LIST

1 INTRODUCTION

1.1 PURPOSE OF DOCUMENT

This procedure is intended to provide the layout, format and control procedures for all internal documents used by Company C. This will conform to the BS 5750 (Part 1) British Standard for Quality Systems.

1.2 SCOPE

Company C has documents that fall into three main areas:

(a) internal documents
(b) internal documents to external specifications
(c) external documents

This procedure covers all the internal documents (a) only, but some of the procedures and controls will also apply to the other documents (b) and (c).

As Company C is not the originator of much of the documentation associated with packages that are sold the control of these documents does not reside within Company C. Other situations are when documents (tenders) have to be produced to a client's specification (i.e. format). In such cases client documentation naming and control procedures should be used.

This procedure does not apply to legal contracts used by Company C.

1.3 TERMS AND ABBREVIATIONS

DW4: DisplayWrite4, a proprietary word processing system available from IBM.

External documents: any documents received by Company C but not controlled by Company C.

Internal documents: any documents produced by Company C for internal and/or external use.

Internal documents to external specifications: any document produced by Company C for internal and/or external use that must conform to a client's specification (i.e. format, control, layout).

PC: personal computer system.

2 APPLICABLE DOCUMENTS AND REFERENCES

BS 5750 (Part 1) British Standards, Quality Systems
QMC 001—Quality Manual
QPC 002—Configuration and Change Control
QPC 017—Non-conformance, Corrective Action and Records
IBM DisplayWrite4 User Manual: Version 3.01
IBM AS/400 DisplayWrite User Reference Manual: Version 1.2

3 TEXT LAYOUT

All internal documents will use decimalised topic headings. However, within the body of any section it is permitted to use either 'roman numerals' (i, ii, iii, etc.), 'alphabetic characters' (a, b, etc.) or asterisks (*) for itemising points.

The number of levels of headings and subheadings should not exceed five. To ensure readability the following should be applied:

(i) the first level of heading should be in upper-case and underlined;
(ii) the remaining levels of headings should be in lower-case and underlined.

The overall layout is:

6   HEADING—FIRST LEVEL
       Text block (if required)...

6.1   Subheading—second level
       Text block...

6.1.1   Subheading—third level
         Text block...

6.1.1.1   Subheading—fourth level
           Text block...

6.1.1.1.1   Subheading—fifth level
             Text block...

The section number is NOT indented and will always reside in the left-most margin. However, the heading is indented to the next available tab stop and also the text block associated with that heading. All headings and text blocks at the same level should start at the same tab stop, thus indenting the main body of the document (but not the section numbers).

If it is necessary to include 'figures' in the body of the text to illustrate a point (e.g. screen format in a user manual) then this is permitted and the figures should be numbered sequentially in the order that they are used. A list of all the figures used should be included after the contents page.

4 FORMAT

4.1 PAGE NUMBERING

All internal documents will be page numbered. The style of the page numbering will depend upon the facility used to generate the document.

On PC-based systems the following style applies as it is easy to implement and control with the facilities of the word processing system, DW4.

The page numbers will appear in the bottom right-hand corner of the document prefixed with the first level section name and the section number. Each section will restart the 'page' number from 1. For example:

FORMAT—5.1

The only exceptions to this rule being the title page, the change history page(s) and the contents page(s). The title page has no page number. The change history page will always have the footer 'Change History—*', where * is the page number. The contents page will always have the footer 'Contents—*'.

On the main computer system, the IBM AS/400, the word processing facility is called DisplayWrite and is not as sophisticated, and the page numbering style has to be simplified.

The page numbers will appear in the bottom right-hand corner of each page of the document, starting with the contents page. The numbering is sequential throughout the document. For example:

Page: 11

The two styles relate to the capabilities of the two word processing facilities used by Company C and will, from time to time, be amended as the facilities of both evolve. It is recognised that the style used for the main computer system will result in the complete reprinting of a document that is changed. For large documents, therefore, the PC style is recommended.

4.2 TITLE PAGE

The title page will correspond to the standard format (see Appendix 1 for example). However, the 'Prepared for:' and 'Contract number:' can be omitted from documents that are created for general use and are not specific to a contract or a person. Any documents intended for internal use only should be marked 'FOR INTERNAL USE ONLY' on the title page and as footer (see Section 5.1).

All documents are uniquely identified by a code. These are discussed fully later.

Details of document status are contained on the title page and are discussed fully later.

4.3 CHANGE HISTORY PAGE

The change history page(s) are used to record and document status, version number and dates of issue. It is a key part of the documentation control process and should always be present in all internal documents (see Appendix 2 for example). The change history page must always refer to the source document that initiates the change.

4.4 CONTENTS PAGE

The contents page(s) will contain references to all section headings and subheadings. The contents page should have as a minimum all headings and the first level of subheadings, the subheadings being indented.

Also included in the contents page(s) should be the list of figures and appendices used within the text in the order in which they occur.

4.5 TEXT BLOCK SPACING

The document will be typed in single line spacing with a double line space between sections (regardless of level), the exception being if contract requirements define otherwise.

Points within a text block can be identified by roman numerals (i, ii, etc.), alphabetic characters (a, b, etc.) or asterisks (*).

4.6 DOCUMENT CONTENTS

The contents of all internal documents should contain the following sections as a minimum:

1. INTRODUCTION
2. APPLICABLE DOCUMENTS AND REFERENCES
3. DETAIL

The remainder of the document should have sections as appropriate.

5 DOCUMENT IDENTIFICATION

All documents produced by Company C will carry a unique identifier. The identifier is formed from three constituents:

(i) an up to three-character mnemonic for the company to which the document relates;

(ii) a two-character mnemonic for the system or procedure to which the document relates;

(iii) a three-digit mnemonic for documents within the system or procedure with incremental numbering in steps of ten.

This document complies with the naming convention and can be used as an example.

5.1 HEADER AND FOOTER

All documents will have a header and a footer formatted as follows:

Header: QPC 008—Documentation Standard and Format

Footer for PC:

'Section heading'—nm

where n is the section number and m is the page number within the section.

Footer for AS/400:

Page: 11

If a document is intended for internal use only then the footer should be extended to include the words 'FOR INTERNAL USE ONLY' on the left-hand side of the footer.

5.2 DOCUMENT STATUS

The status of a document determines the level of control that is used with that document. The status is identified on the title page only and can have one of the following values:

    (i) U—for uncontrolled
    (ii) C—for controlled
    (iii) F—for frozen

An uncontrolled document can be changed by the originator/compiler of the document and need not be referred for approval.

A controlled document can be changed only after an internal management/development/technical review and corresponding approval.

A frozen document can be changed only after an internal management/ development/technical review and consultation with the 'owner'. This in the majority of cases will be the client for which the document has been produced.

5.3 DOCUMENT ISSUE

The issue of a document will always start at 01.00 until it is first published. Thereafter any modifications/changes to the document that do not result in the need to print the complete document again will result in an increment to the lower level issue number (e.g. 01.01). The reason for printing the complete document again might be for ease of distribution due to 'many' changes and this will result in an increase in the higher level issue number (e.g. 02.00). When the higher level issue number is incremented then the lower level is always reset to zero.

5.4 CONTROLLING CHANGES

All documents in a controlled or frozen state can only be changed by the provision of a 'new' change history page(s), relating to the changed pages or sections of the document. Only in the case of controlled and frozen documents must the changes be reviewed and authorised.

6 DOCUMENT CONTROL

All documents are controlled from the first publication of the document. To ensure that unique document identifiers are used, a central document registrar is used. (See the appendices for example forms and refer to QPC 002—Configuration and Change Control for control details.)

APPENDIX 1: TITLE PAGE

COMPANY C
QUALITY PROCEDURE QPC 008
DOCUMENTATION FORMAT AND STANDARD

Compiled by...

Checked by..

Approved by..

Date...

APPENDIX 2: CHANGE HISTORY

CHANGE HISTORY

Document status/version	Date issued	Number of pages	Changed pages	Comment/error reference
U–01.00	16/02/89	n/a	n/a	First release
U–01.01	17/02/89	5	Appendices	Second release
C–01.01	01/07/89	2	Title/Change History	009
C–01.02	11/07/89	2	5.1–5.2	0159

APPENDIX 3: COMPANY IDENTIFIER LOG

Company Identifier Log	
Id.	*Company name and address*
CCS	*xxxxx*
HOS	*yyyyy*
THN	*zzzzz*

APPENDIX 4: SYSTEM IDENTIFIER LOG

System Identifier Log	
Company Id.: THN	*Company name*: THN
Id.	*System name and brief description*
PS	Product system— A system designed for THN and developed on Synon/2 to control the publication process from conception to final print.
CS	Commercial System—A system designed by Company C as a package and modified to meet the requirements of THN. The system controls the calculation and payment of accounts.

387

APPENDIX 5: DOCUMENT IDENTIFIER LOG

Document Identifier Log				
Company Id.: THN		Company name: THN		
System Id.: PS		System name: Product System		
Doc. No.	Document name	Alloc. date	Originator	Publish date
010	System description	01/12/88	A. N. Other	10/02/89

APPENDIX 6: DOCUMENT DISTRIBUTION LIST

Document Distribution List			
Company Id.: THN		Company name: THN	
System Id.: PS		System name: Product System	
Document Id.: 010		Document name: System description	
Doc. issue	No. of copies	Sent to: (name and address)	Date sent
01.00	1	AA	12/02/89
	1	BB	12/02/89
01.01	3	CC	17/02/89
	1	DD	17/09/89

COMPANY C
QUALITY PROCEDURE QPC 009
PURCHASING AND SUPPLIERS

Compiled by ....

Checked by ....

Approved by ....

Date ....

CHANGE HISTORY PAGE

Document status	*Date issued*	*Number of pages*	*Changed pages*	*Change/ defect no.*
XX	XX	XX	XX	XX

CONTENTS LIST

1 INTRODUCTION

1.1 PURPOSE OF DOCUMENT

This procedure is intended to provide the control mechanisms, reviews and audits of the purchase of goods and/or services by Company C. This will conform to the BS 5750 (Part 1) British Standard for Quality Systems.

1.2 SCOPE

Company C has suppliers that fall into three main areas:

   (i) suppliers of services, consumables and products for internal use only;
  (ii) suppliers of services and products that are passed on to customers;
 (iii) client supplied/loaned products supplied for development.

This procedure shall apply when assessing all suppliers and in the control of purchasing, goods inward and returns to suppliers.

1.3 TERMS AND ABBREVIATIONS

None.

2 APPLICABLE DOCUMENTS AND REFERENCES

BS 5750 (Part 1) British Standards, Quality Systems
QMC 001—Quality Manual
QPC 008—Document Format and Standard
QPC 004—Software Development
QPC 006—Software Review
QPC 005—Software Design Standards
QPC 002—Configuration and Change Control
QPS 007—Subcontracted Software
QPC 015—Test
QPC 017—Non-conformance, Corrective Action and Records
QPC 019—Internal Quality Audit

3 SUPPLIER ASSESSMENT

3.1 ASSESSMENT METHOD

The assessment method of assessing and approving a potential supplier is:

(i) by evaluating the information returned on the supplier questionnaire (see Appendix 1);
(ii) by reputation and previous service; and
(iii) by accepting the approval of other companies with known quality
standards.

The type and extent of any evaluation shall be dependent upon the
nature of the products or services to be provided and the degree of previous
experience with that supplier.

Suppliers of major products or services shall be subject to an initial
quality audit to establish their quality assurance capability. They will be
monitored by review every year, or more frequently should any problem
with quality arise.

The result of the review shall be reported on the supplier quality
assurance rating form (see Appendix 2), which will be filed in the supplier
quality folder. The supplier quality assurance form will show the category
of product or service reviewed, and record the assessment as one of

(i) fully approved
(ii) partially approved:approved for a specific product or service
(iii) unsuitable as a supplier

3.2 LIST OF APPROVED SUPPLIERS

A list of approved suppliers will be held in the supplier folder. The folder
will be divided into three sections:

(i) fully approved suppliers
(ii) partially approved suppliers
(iii) unsuitable suppliers

From time to time the unsuitable suppliers list will be purged and any
suppliers with no activity for a period in excess of 18 months will be
removed from the list and placed in the redundant supplier file, where
details are kept for three years.

An unsuitable supplier may request a further review, and providing a director of Company C approves such a review it may proceed.

Suppliers either fully or partially approved may be reviewed at any time, other than the yearly review at the discretion of the technical director, or if the performance of the supplier falls below 60% as measured by the performance category (see Section 3.3).

3.3 PERFORMANCE CATEGORY

The performance of a supplier is determined by the percentage of time the order is fulfilled correctly and delivery is on time:

$$\text{Performance} = \left[\left(\frac{\text{Correctly fulfilled orders}}{\text{Total orders}}\right) + 100\right]\%$$

The performance is calculated on a quarterly basis and results kept in the supplier performance file.

4 PURCHASING

4.1 PURCHASE ORDERING

Any item to be purchased, be it a product or a service, must have the necessary approval. The approval can take two forms:

(i) letter of authorisation from a customer
(ii) purchase order signed by a director of Company C

The letter of authorisation from a customer will allow a purchase order to be raised by Company C on behalf of the customer. This applies to all products not held directly by Company C (e.g. computer hardware and software).

The purchase order will contain details of delivery address, invoice address, product and/or service required, required delivery date or start and completion date in the event of a service, and the cost of the product and/or service (see Appendix 3). Once a purchase order has been approved it is allocated the next available purchase order number and the details recorded on the purchase order control list. Then one copy of the purchase order is sent to the respective supplier and the second copy is filed in the outstanding purchase order folder.

4.2 PURCHASE ORDER TRACKING

The purchase order control list is reviewed by the office administrator on a weekly basis and a report compiled of outstanding purchase orders. The report should detail the original delivery date and the latest expected delivery date, and be in purchase order number sequence.

All outstanding purchase orders should be chased to ensure that the product and/or service will be delivered on time. The criticality of the product and/or service will determine the extent of the chasing for the product and/or service. The minimum should be that the mid-point between the purchase order date and the required delivery date should be the time when the chasing occurs. All chasing should be recorded, attached to the purchase order and filed in the outstanding purchase order folder. If a slippage in the delivery date is expected, the purchase order control list is updated.

5 PRODUCT INWARD, REVIEW AND RETURNS TO SUPPLIER

5.1 PRODUCT RECEIPT

All products received by Company C will be checked for quantity and quality. If a goods received note has to be signed then this will not be done until all the product has been checked. A carrier's note may be signed with the words 'goods not seen' printed across the carrier's note.

If the product is delivered directly to the customer then the office administrator must check with the customer that the product has arrived and is suitable for its intended purpose.

5.2 PRODUCT REVIEW

All products should be reviewed to ensure that they are suitable for the intended purpose and that the quantity specified on the purchase order matches that received. If this is the case the purchase order is marked as complete and the product stored in a suitable location.

If the product is damaged and/or short on quantity not specified on the goods received note from the supplier, the supplier must be informed within 24 h and corrective actions agreed in writing.

If the product is over-delivered, again the supplier must be informed within 24 h and corrective actions agreed in writing. The excess product is marked accordingly and stored in a secure location away from the normal storage location of that product.

The review to establish if the product is suitable for that which it was intended will depend upon the nature of the product and if the product has previously been fully audited. However, if a product is new to Company C then a full review and audit must take place and time must be allowed for this, even to the extent of the supplier providing inspection copies of the product.

5.3 RETURNS TO SUPPLIER

All returns to suppliers must be agreed in writing between Company C and the supplier prior to the product leaving the control of Company C. Within the written agreement should be detailed the responsibility for the product during shipment (i.e. insurance, carrier charges, etc.).

6 SERVICE INWARD AND REVIEW

6.1 SERVICE INWARD

Any service provided by a supplier to Company C should be of high quality. The service should be provided as defined in the purchase order and/or the terms and conditions of reference used for that service; such terms should be provided by Company C.

6.2 SERVICE REVIEW

The service will be reviewed in the same way in which services provided by Company C are reviewed. This required reference to virtually all of the procedures used to control quality by Company C.

6.3 UNSATISFACTORY SERVICE

If after a review the service provided by a supplier is found to be unsatisfactory then that supplier must be informed immediately in writing. A meeting should be arranged with the supplier as soon as possible to discuss

    (i) corrective action
    (ii) possible compensation

If the service is being provided to a client of Company C then that client must also be informed of the problem and an agreed action plan implemented to minimise the risk to the client.

7 CLIENT SUPPLIED/LOANED EQUIPMENT

Where client supplied/loaned equipment is used on a project the project or quality manager will be responsible for keeping a record of all materials (normally computer hardware or test equipment) provided on loan for use on Company C's premises. In addition, the manager will be responsible for ensuring that the following aspects are agreed with the customer and that the agreed conditions are included in the quality plan or other project document. These include:

—responsibility for providing and paying for maintenance to the equipment;
—responsibility for providing and paying for insurance;
—provision of and payment for consumables (media, paper, etc.);
—Company C acceptance of equipment;
—acceptance criteria;
—acceptance signatures.

APPENDIX 1: SUPPLIER QUESTIONNAIRE

1. The results of the supplier evaluation shall be reported to the technical director.
2. Does the contractor have a quality system that meets the requirements of BS 5750 (Parts 1, 2 and 3)?
3. (a) Does the contractor have an individual responsible for quality/ inspection?
   (b) Is he independent of other functions (i.e. manufacturing and production)?
   (c) Does he have the authority to enable him to resolve all quality matters?
4. (a) Does the contractor have adequately documented procedures for design (i.e. quality manual, written software design standards)?
   (b) If not, how does he ensure vital review functions are carried out?

(c) Does the contractor have a system for the formal updating and revision of standards and work instructions vital to the quality of the products?

(d) Does the contractor maintain review and test records for a prescribed period?

(e) Do the records verify that essential test/performance have been carried out?

(f) Does the contractor have a recognised corrective action procedure?

5. Does the purchasing documentation contain clear descriptions of the products required and any necessary supporting documentation?

6. Does the contractor have procedures which ensure that only the latest applicable technical data and/or specifications are used?

7. (a) Does the contractor's system ensure that only acceptable items are used for manufacture or distribution?

(b) Are non-conforming items at the goods inwards, assembly test and inspection stages identified with a rejection tag and segregated?

(c) Are records of non-conforming items and their disposition maintained?

8. Does the contractor have a procedure for the return of non-conforming goods (i.e. rejection notes, rejection tag)?

9. (a) Are all inspection and test devices in a known state of calibration?

(b) Is documentary evidence available?

(c) By what method does the contractor determine the frequency of calibration of inspection and test devices?

10. (a) Does the contractor keep inspection and test records?

(b) If so, for how long?

11. Does the contractor have a review and evaluation system to ensure all quality functions are kept up to date?

12. Is there any final inspection of goods carried out before despatch to ensure conformity with order?

13. What methods (e.g. static analysis) are used for the validation of products?

14. What formal methods (e.g. VDM, Z, OBJ) are used?

Appendix 2 follows

APPENDIX 2: SUPPLIER QUALITY ASSURANCE RATING FORM

Supplier Quality Assurance Rating Form
Supplier name: Supplier id.: Supplier address:
Date of rating: / /   Rating: Good/Average/Poor
Q1. Are quality control procedures in place? (Y/N) Q2. Do the quality control procedures comply with BS 5750? (Y/N) Q3. Is the supplier to provide limited product/service? (Y/N) Q4. If yes, what are the products/services?
Completed by: Date: / / Rating: fully approved/partially approved/unsuitable Comment:

APPENDIX 3: PURCHASE ORDER FORM

Purchase Order		
Purchase Order No.: Date raised: / /		
Supplier Name: Supplier id.: Supplier Address:		
Required delivery date: / /		
Purchase order raised by: Date......................: / /		
Purchase order details:	Quantity	Price
	VAT	
	TOTAL	
Delivery address: Special instructions:		
Authorisation Signature...................................... : Printed name................................ : Date .... :		

COMPANY C
QUALITY PROCEDURE QPC 011
GOODS INWARDS, INSPECTION AND STORES

Compiled by...

Checked by...

Approved by..

Date...

CHANGE HISTORY PAGE

Document status	Date issued	Number of pages	Changed pages	Change/ defect no.
XX	XX	XX	XX	XX

CONTENTS LIST

1 INTRODUCTION

1.1 PURPOSE OF DOCUMENT

To ensure that goods received meet the conditions of the purchase order and that records of incoming goods are kept.

1.2 SCOPE

All items procured by Company C.

1.3 TERMS AND ABBREVIATIONS

Complete shipment: all the items specified on a purchase order when they have been accepted as conforming.

Partial shipment: a shipment whereby some of the items have been received and conform, and whereby the vendor has indicated on the despatch note that the missing items will follow.

Incorrect shipment: a delivery that contains either non-conforming items or not all of the items.

Goods received note: a document, raised on receipt of the goods, which is passed to administration.

2 APPLICABLE DOCUMENTS AND REFERENCES

QMC 001—Quality Manual
QPC 017—Non-conformance, Corrective Action and Records
QPC 009—Purchasing and Suppliers

3 RECEIVING PROCEDURE

When items are received at Company C, administration will direct them to the appropriate project. A member of the project team (normally the originator of the purchase order) will inspect the items for conformance. Where necessary he will delegate specialist tests or inspections.

4 VISUAL INSPECTION

4.1 THE INSPECTION

On arrival of the items at Company C the member of the project (see Section 3 above) will:

—refer to the purchase order;
—count and inspect items for damage or wear;
—note discrepancies on the goods received note;
—if alternatives have been substituted, check the acceptability thereof;
—carry out tests as necessary, consulting other staff as appropriate;
—check the issue status of software media;
—check the labelling and release documentation.

4.2 RECORDING VISUAL INSPECTION

Details will be recorded on the goods received note and details of any discrepancies will be described. The goods received note will then be signed and passed to administration.

4.3 FURTHER INSPECTIONS

Where specialist tests are needed to verify the incoming items then the technical director will be consulted for support.

5 NON-COMPLIANT GOODS

If goods are defective the finance director will be informed and will contact the vendor. Deficiencies or non-conformances must be resolved speedily so as not to adversely affect schedules.

If the vendor cannot remedy the problem, alternatives will be sought, which may include:

—accepting the items for in-house rectification;
—using an alternative vendor.

6 STORAGE

All products under development will be kept in the correct environmental conditions, whether in use or otherwise. All magnetic media will be stored in controlled conditions.

COMPANY C

QUALITY PROCEDURE QPC 015

TEST

Compiled by...

Checked by...

Approved by ....

Date...

CHANGE HISTORY PAGE

Document status	*Date issued*	*Number of pages*	*Changed pages*	*Change/ defect no.*
XX	XX	XX	XX	XX

CONTENTS LIST

1 INTRODUCTION

1.1 PURPOSE OF DOCUMENT

The purpose of this procedure is to provide guidance on the testing of all products provided by Company C.

1.2 SCOPE

The scope is to cover all products supplied by Company C. Calibration does not apply.

1.3 TERMS AND ABBREVIATIONS

DFU: data file utility, provided by IBM as a system facility on the IBM AS/400 and the IBM S/36, and allows the creation and amendment of records in data files.

Query: query, provided by IBM as a system facility that allows for the printing of information from a data file and also allows selection and sort criteria to be applied to the listing.

2 APPLICABLE DOCUMENTS AND REFERENCES

BS 5750 (Part 1) British Standards, Quality Systems
QMC 001—Quality Manual
QPC 002—Configuration and Change Control
QPC 005—Software Design Standards
QPC 004—Software Development
QPC 017—Non-conformance, Corrective Action and Records
QPC 006—Software Review

3 SOFTWARE TESTING STRATEGY

3.1 INTRODUCTION

The purpose of software testing is to establish the 'correctness' of that software for the function for which it is intended. The products provided by Company C fall into three main categories, namely

   (a) externally controlled software packages

(b) internally developed software packages

(c) internally developed software packages using 4GLs (fourth generation languages)

The mechanisms available for testing also vary. In the case of external software packages no source code or system specification details are available and therefore no static or structural testing can be performed. In the case of 4GLs the level at which modifications are allowed is very high and many standard functions are provided directly from a data model; again, in such a case the testing procedure is modified.

3.2 FUNCTION TESTING

Functional testing is a mechanism whereby the requirement to be satisfied of a software unit is known and both the input and the output can be monitored for correctness. Normally the document used for functional testing is the system description.

To ensure that a complete function test is performed a function test plan should be produced as the software units are developed from the system description; the test plan can be a separate document or part of the system description.

However, a one-to-one test plan and software unit should always exist. The function test plan should contain testing procedures for *all* software units within a system. Each software unit should be tested for valid and invalid input, and the expected result of valid and invalid input should be known prior to any test commencing. An example of a function test plan for a single software unit is contained in Appendix 1.

A true function test plan will be exhaustive in testing every conceivable permutation of input. However, this is not possible in real terms and therefore some generalisation has to be made in relation to the contents of a functional test plan. The exact level of testing will be determined by commercial constraints and each case must be judged separately.

Ideally an independent tester should be appointed to perform the functional testing and this tester should never be the developer of the software unit.

3.3 STRUCTURAL TESTING

Structural testing involves the use of 'bench' testing or 'dry run' testing and the tester must have access to the source code of the software unit for it to be performed.

The purpose of the structural test is to establish, for example, if any infinite loops are present in the software unit and also whether the software unit complies with the checks and requirements required within the system description.

Example of an infinite loop:

```
SET CONDITION INVALID
DO UNTIL CONDITION VALID
   PROCESS INFORMATION
   CHECK INFORMATION
END DO
```

The above example would be an infinite loop unless the condition was set to valid within the DO loop.

The level of the structural test depends upon the programming language used. If a 4GL is used then many checks and validations are implicit and cannot be checked except by the use of a functional test, described above. The 4GL has in-built processing logic that allows the creation and modification of a type of software unit. Within the software unit it is only possible for the programmer and analyst to specify additional actions at key points which do not affect the main logic flow of the software unit.

The structural test should not have a formal plan of action and is normally conducted by the programmer of the software unit. If a software unit is complex then it may be the project leader who would conduct the structural test.

Ideally an independent tester should be appointed to perform the functional testing and should never be the developer of the software unit.

4 FUNCTIONAL TESTING STANDARDS

4.1 PROGRAM TESTING

It is imperative that every program written receives a full and complete test of its capabilities. This should be carried out immediately after the program has been coded, or as near as practicably possible, by the person who coded the program.

The object of program testing is to ensure that every line of coding performs in accordance with the program specification and complies to the programming standards of Company C.

Broadly speaking, there are two types of program:

(i) on-line or interactive
(ii) batch or off-line

4.1.1 On-line Programs

The distinguishing feature of these programs is that they use a screen file in order to communicate with a workstation (terminal).

Test the following on every screen:

(i) Screen contents: ensure that all the screen contents are correct in

  (a) position
  (b) content (spelling, etc.)
  (c) presentation (highlight, underline, etc.)

(ii) Output screen variables: ensure that every output field is correct in

  (a) position
  (b) content
  (c) presentation
  (d) length

(iii) Input/output screen variables: ensure that every input/output field is correct in

  (a) position
  (b) content
  (c) presentation
  (d) length
  (e) validation and cross-reference

(iv) Cancel/end: ensure that the screen allows cancellation or ending of the function and that this works correctly.
(v) Confirm prompt: if the screen includes a confirm prompt that the correct default answer is displayed and that the confirmation works correctly.
(vi) Inhibited keys: ensure that any invalid keys are inhibited on the screen.
(vii) Scrolling screens: ensure that the screen will correctly handle empty screens and multiple screens, and no more records to display on screen. Also ensure that scrolling works in a forward and backward direction.

409

Test the processing sequence of screens as follows:

(i) Test that the screens are displayed and processed in the correct sequence. Check that this is true in all functions, i.e. entry, amendment, delete, etc.

(ii) Check that when cancel/end is selected the correct screen is displayed and processed.

Print all files that have been attached to the program and check the following:

(i) Check that any files that are input only have not been altered.

(ii) Check that the record length and format of all amended or created records is correct.

(iii) Check that each newly created record is correct (field by field). Especially check that fields not input have been correctly initialised: numeric fields with zeros; packed fields with packed zeros; alphanumeric fields with spaces.

(iv) Check that each record updated has been updated correctly (field by field).

4.1.2 Batch Programs

Normally speaking, a batch program consists of input, input/output and output files.

Before testing a batch program it will be necessary to set up input and input/output files in such a fashion as to test every part of the coded program. In order to do this it is necessary to produce a plan of the possible permutations and combinations of data, and from this draw up the data that will be needed to reside in the files.

It is preferable to phase the testing of these programs, beginning with the simplest test and progressing to the most complex combinations of data. Always test for the correct processing of empty files.

After each run of the program, print all files that have been attached to the program and check the following:

(i) Check that any files that are input only have not been altered.

(ii) Check that the record length of all amended or created records is correct.

(iii) Check that each newly created record is correct (field by field). Especially check that fields not input have been correctly initialised.

(iv) Check that each record updated has been updated correctly and that fields that should not have been altered have remained the same.

410

Additionally, for each printed report check the following:

  (i) Check page headings, in particular page numbering, date and date format.

 (ii) Check each printed line, in particular correctness and format of each field, zero suppression and negative numbers.

(iii) All control breaks, totals and subtotals.

(iv) Especially test that the program correctly handles a full page, even if it means setting up a large amount of test data.

 (v) Check that the '*END OF REPORT*' legend is printed correctly.

Refer to Appendix 2 for program testing checklists.

4.2 TEST PLAN

To ensure a program receives a full and complete test in the most efficient manner it is necessary to design a test plan before testing commences, to adhere to the plan, and to anticipate and record the results.

4.2.1 Design of Test Plan

Many programmers simply sit down in front of a terminal and think: 'What shall I do now? I'll try that.' This is not acceptable and leads to numerous errors, sometimes of the most obvious kind, not being detected during a program test.

A programmer should never test anything until he or she is confident that the program is going to work, and that he or she knows exactly what the current test is designed to achieve and the type of errors that are likely to occur.

Put down on paper the areas that are being tested and circumstances under which they can arise. Note down the data you are going to use: codes, descriptions, commands or whatever. Testing is a laborious task with no short-cuts, and preparation is essential.

The test plan should be formatted, neat, consistent (i.e. check similar processes in the same way for each test), re-usable and, most of all, comprehensive.

The test plan should contain details of the files used and these files should always be saved for further tests if required; only when the program is fully tested and reviewed should test files be removed, and only then if they are not required for further processes.

411

4.2.2 Adhere to the Plan

If the plan is not adhered to it becomes useless. Preparing the data to be put through gives the best chance of ensuring all circumstances are covered. When bugs/errors are traced the same data can and should be re-applied and exactly the same tests undertaken. This enables one to easily identify that amendments have been applied correctly.

4.2.3 Anticipate the Results

Whenever a test is undertaken the programmer should know what results are expected. This makes it easier to identify whether the test has been successful or otherwise and ensures greater thought is given to the coding, resulting in a better understanding of what the program is attempting to achieve.

It is a common fault that programmers feed data into a program and then attempt to justify the results that the test has given. This is entirely the wrong approach to testing and demonstrates a lack of discipline; it is also an inefficient use of the programmer's time. Only when the results produced from a test are different to what is anticipated should there be investigation into the cause.

4.2.4 Record the Results

Keep notes of everything that occurs in testing, including file prints, reports and screen prints. It is not necessary to provide a listing of the program. The notes should be kept in the program folder started at the same time as work on the program and will also include the program specification and other notes.

Reference to the program folder at a later date will ensure that future test plans are produced quickly and efficiently, and that common, but not necessarily obvious, errors can be more easily identified.

If all results and pertinent details are retained they can be used to demonstrate that each test has actually been performed. It will ensure that all areas receive testing and that specific areas are not tested more than once. If a programmer has to stop in the middle of a test then when recommencing he or she may start in the wrong place and finish up doing too little or too much testing.

It is often useful to be able to relate back to a set of results should, say, an error occur in the live running of the program.

412

These test plan guidelines apply equally to both the simplest and the most complex programs.

4.3 GENERAL POINTS AND COMMON MISTAKES

It must be stressed once again that there are no short-cuts or easy options when undertaking program testing, and that good and thorough testing requires concentration and discipline.

All paths, circumstances and conditions should be tested but do not concentrate on the obscure to the exclusion of the obvious.

Be efficient with both time and resources. Always continue to test until you can get no further. If one function of the program fails, continue the test with another. Do not find an error, edit, re-compile, find the next error, edit, re-compile. This not only wastes time but hinders others who may be waiting for a terminal or use of the compiler or whatever.

All functions of the program should be independently tested, even when using mostly common code. Simply because all fields operate correctly under circumstances of addition does not necessarily mean they will work under amendment.

It is essential that programmers do not presume that things are going to work. Errors often occur when one section of code has been copied to undertake a similar job elsewhere. Simply because the first process is going to work does not necessarily mean that the copied process is going to work. If a separate section of code is used then that code should be thoroughly and independently tested.

A very important aspect that is almost always overlooked, particularly in large interactive systems, is the area of file lockout and subsequent processing. Lockout situations must always be tested because in live situations programs often run in parallel, and it is this area that is likely to cause more operational problems than any other.

Take-on and one-off programs should be tested just as thoroughly as everyday programs, even though they may only be executed once. All programs, irrespective of their end objective or frequency of use, have to operate without failure or error.

The simplest of changes sometimes necessitates thorough retesting, even in the areas not immediately related to the amendment made. Far too frequently a change is made to cure one problem only for it to generate another problem.

It is important to be aware that a change to a program which supplies data for another program may necessitate the other program being

413

retested, even though it itself has not been amended. In cases such as these the responsibility of the individual to retest a particular program may be unclear. This should not result in testing being neglected and there will always be a project leader who will clarify the situation if necessary. Never presume in such cases that the responsibility lies elsewhere.

4.4 PROGRAM TESTING AND PROGRAM SPECIFICATION

A program specification can never be considered a perfectly accurate, non-ambiguous document, and as such there will always be details which have been omitted or are superfluous, even misleading or simply incorrect. It is the responsibility of the programmer to be aware of this and to bring any such cases to the attention of the project leader.

Although the above comment is made, all possible effort is used to ensure accurate and precise program specifications via the use of the latest methods and techniques in system design methodology. In the same way, often the program specification will not state the obvious and the programmer must act responsibly in respect of these 'omissions' by ensuring that not only does the program perform to specification but also that it fulfils the spirit of what is required and it is operationally presentable.

Certain details often not spelled out in the specifications are nonetheless the responsibility of the programmer and should be coded and tested (for consistency, for example) both within the program and across programs, of error handling and messages, use of field emphasis (highlight, underline, etc.) and processing sequences. Such processes are detailed within the technical reference guide started at the beginning of each project.

4.5 SYSTEM TESTING

A program cannot be tested in total isolation. It must always be remembered that a program is simply a component of a system. It will often be dependent on other programs and have other programs dependent upon it. Ensuring that dependency is fully tested requires cooperation between the programmers involved and a joint discussion of results. A system test checklist is provided in Appendix 3.

It is the responsibility of the project leader to ensure that system dependency is correctly stated and handled within a system test plan. In addition, during the development of the system, programs are created in the correct sequence so as not to prevent successful program testing.

Although initial system testing should be performed by the members of the development team, the final system testing should be performed by an independent party who has the responsibility for the final system testing and any minor modifications/error corrections that may be required at this stage. It is therefore vital that the results of the various program tests and previous system tests are maintained and available for use during the final testing stage.

On completion of the final system test, the system can be reviewed and changed to 'control' status. As any system may be divided into many sub-systems it is possible that this review procedure occurs many times during the life of a project as each sub-system is made available.

4.6 TEST DATA

The creation and use of test data is an important part of program and system testing, and is probably the most neglected element of program and system testing.

4.6.1 Quantity

Use plenty of data; its entry may be time-consuming but it is efficient use of that time—it may also assist the testing of the entry programs within the system under test. It is common to come across obvious errors such as page throws or scrolling screens which have not been properly tested simply because the programmer did not enter a sufficient quantity of data to cause these situations to occur.

Should a program require large amounts of data in order that testing can be thoroughly accomplished it is the responsibility of that program's author to ensure that sufficient data exists on the files. The fact that it is not his program that creates or enters such data is no excuse for insufficient volumes to be present.

4.6.2 Quality

Ensure the data used is diverse and complete. Use a variety of different values, e.g. in numeric fields have data with decimal positions, no decimal positions, zero, maximum valid values, minimum valid values, etc. In alphanumeric fields ensure that fields are entered to their full length. Structure data in such a manner that is easy to spot whether truncation is taking place; be aware of numeric rounding and overflow problems.

Frequently records are created to add volume to the test system only for

415

the majority of their non-mandatory fields to be left empty. The more records there are on file, the greater the number of entered fields these records contain, and the more varied in length and content these fields are, the greater the opportunity to track down cosmetics and errors of presentation.

Above all, the data should be meaningful and reflect the purpose for which the system or program is intended.

4.6.3 Accuracy

In the normal course of system development and testing, data becomes corrupted on files. Whenever these situations arise the corrupt data must be responsibly removed or amended. With all the utilities available to us there is never any excuse for corrupt data remaining on file.

It is not unusual for a programmer to explain unexpected test results on the fact that the data in the test was corrupt, only for it to transpire that although the data may have been corrupt this was not actually the reason for the results, and as a consequence program errors remain undetected and uncorrected.

It is the responsibility of each individual programmer to ensure the accuracy of the data he or she maintains on a shared test system.

4.6.4 Exclusivity

It is impossible to accurately test what is happening to an individual record if several other programmers are updating it within the duration of the test. Therefore use exclusive data when necessary, either by maintaining records exclusive to you within a shared test system or, if this is not possible or practicable, by creating a controlled personal test system.

If exclusive test systems are utilised it should still be remembered that the live system will not be so sterile and that dependency and relationships must also be considered.

4.6.5 File Prints

It is a strong temptation, especially on interactive systems, to judge results purely by reports and screens. Although these must be looked at, they are not a substitute for prints of the data files.

There are often occasions when corrupt data cannot be located or identified without recourse to the physical data, for example system data (as opposed to entered data) and initialised fields on addition of records. There may even be occasions when fields are maintained but not intended for use until a later phase of development, and therefore their accuracy cannot be confirmed by any method other than investigation of the appropriate data files.

All programmers must be proficient with the use of the HEX system of numbering. All packed numeric fields are stored in data files as HEX fields.

4.6.6 Individual Record Amendment

Ensure that when data is entered or amended that the results are checked at the correct time, performing, if necessary, tests on separate records, e.g.

(i) test 1 on record A
(ii) test 2 on record B
(iii) tests 1 and 2 on record C

Performing test 1 followed by test 2 on the same record, without checking the intermediate results, will not always be adequate or acceptable.

5 SOFTWARE PACKAGE TESTING

A software package obtained from an external supplier is normally obtained without access to the system description or source code. It is therefore necessary to devise a functional test plan based upon the user documentation. All the procedures and guidelines detailed in Section 4 should be applied as appropriate.

6 SOFTWARE DEVELOPMENT TESTING

Any software developed by Company C will either use normal high level languages, such as COBOL or RPG III, or a 4GL, such as Synon/2 or IPG400.

Appendix 1 follows

417

APPENDIX 1: FUNCTIONAL TEST PLAN

System name: Sales ledger
Program name: Customer file maintenance

1. Creation of new customer record.
2. Creation of duplicate customer record (should not be permitted).
3. Amendment of customer record.
4. Amendment of customer record that does not exist.
5. Reference within function to invalid codes (table file) should not be permitted.
6. Aborting of program during customer record creation should result in no new customer record.
7. Are all fields being updated correctly for amendment and creation?
8. Aborting of program during amendment of customer record should result in no change to record details.

APPENDIX 2: PROGRAM TESTING CHECKLIST

ON-LINE PROGRAMS—SCREEN

- format correct
- standard header and footer correct
- constants correct
- output variables correct
- input/output variables correct
- inhibited keys prevented
- scrolling correct (backwards and forwards)
- validation correct

ON-LINE AND BATCH PROGRAMS—FILES

- input files unchanged
- input/output files correctly updated
- output files correctly updated

ON-LINE AND BATCH PROGRAMS—GENERAL

- first time through processing
- empty input files
- invalid data entry

BATCH PROGRAMS—GENERAL

- report headings correct
- break points correct
- subtotals and totals correct
- page break processing correct
- '*END OF REPORT*' legend correct

APPENDIX 3: SYSTEM TESTING CHECKLIST

FUNCTIONS

- all individual functions correctly program tested
- system test plan completed
- data flow correct through system

COMPANY C
QUALITY PROCEDURE QPC 017
NON-CONFORMANCE, CORRECTIVE ACTION AND RECORDS

Compiled by...

Checked by...

Approved by ....

Date...

CHANGE HISTORY PAGE

Document status	Date issued	Number of pages	Changed pages	Change/ defect no.
XX	XX	XX	XX	XX

CONTENTS LIST

1 INTRODUCTION

1.1 PURPOSE OF DOCUMENT

This document outlines the mechanism for the control of the manual error recording system used by Company C.

1.2 SCOPE

Company C has a need to record all errors reported to the company in relation to all the products and services supplied. The errors fall into four main categories:

(a) hardware error
(b) operating software error
(c) package software error
(d) bespoke software error

This document covers the mechanisms to be used in all the above error categories. The reason for the error is not relevant at this stage.

1.3 TERMS AND ABBREVIATIONS

Hardware error: any error that results in the failure of a hardware component of the system used by a client.

Operating software error: any error that is due to the proprietary operating software that is supplied with the hardware and allows the hardware to function as a computer system in the basic sense.

Package software error: any error that is due to deficiencies in software written and sold as a working system. Such a package may not be under the direct control of Company C, e.g. where sold under licence.

Bespoke software error: any error that is due to deficiencies in software written and sold by Company C.

2 APPLICABLE DOCUMENTS AND REFERENCES

BS 5750 (Part 1) British Standards, Quality Systems
QMC 001—Quality Manual
QPC 002—Configuration and Change Control
QPC 006—Software Review

3 PROCEDURE FOR ERROR LOGGING

3.1 INITIAL CONTACT

The initial contact is made either by the client to Company C or the error is identified by Company C; this can be by any means of communication but normally by telephone.

3.2 ERROR LOGGING

The date and time of contact, together with the client name, is recorded on the error report and a new error number obtained from the error log, on which the same information is recorded (see Appendices 1 and 2 for samples).

The error log is used to produce a summary of errors during a week, and controls the allocation of error numbers.

The allocation of an error number to a client enables the same error to be tracked (see Section 4.3).

3.3 EXCEPTIONS

As client sites are often visited by Company C staff, situations will occur where errors are reported by the client directly to those members of staff. In this instance all staff involved in site visits shall carry blank error reports that can be completed on site.

Error numbers can be allocated when the staff member returns to the office if the error is not successfully resolved on site, or the Help Line service should be contacted to allocate an error number if the problem cannot be resolved on site.

4 PROCEDURE FOR ERROR TRACKING

4.1 INITIAL CONTACT

The initial contact stage is used to define the problem in sufficient detail to enable the information to be passed to a technician if required. The format of the error report is designed to facilitate this information gathering.

During the initial contact stage the following information should be recorded:

(a) Date and time of contact.
(b) Client company name.
(c) Who reported the error and position within the client company.
(d) Telephone number of client and extension to contact the person who reported the error.
(e) Who logged the error within Company C.
(f) Type of hardware the error occurred on and the model number if known.
(g) Device in error if a hardware error.
(h) Operating system and associated version number if known.
(i) Package in use.
(j) Classification of error.
(k) Description of error.
(l) When the problem occurred and whether it was on-line or in batch.
(m) Check if power is available to the devices.
(n) Priority of the error.
(o) 'Solved box', which should be ticked when the error is solved and is positioned on the form for quick reference whilst searching for outstanding errors.

4.2 CLASSIFICATION OF ERROR

Errors are classified as follows:

Hardware: any error that prevents the hardware from physically working.

Operating system: any error that prevents the computer system operating in its most basic mode.

Package software: any error reported by the client in products designated to the packaged software.

User training: any error that is due to the client not understanding the workings of the computer system in sufficient detail and causing a problem to develop.

User manual: any error resulting from an instruction provided in a user manual.

Other: any other reported error with associated details.

424

4.3 EVALUATION AND SOLUTION

Once the details of the error have been recorded then an evaluation and solution can be found.

The error is passed to a person within Company C, who is then responsible for the evaluation and solution to the problem. The date and time of the passing of the problem is recorded on the error report.

If the solution cannot be found within a reasonable period then it must be referred to a second party. The date and time of the referral is recorded on the error report.

The initial person responsible for the solution is then responsible for ensuring that the second referral is answered in a reasonable period. The date and time of the answer is recorded on the error report.

Once the error has been fully resolved it is cleared and the date and time recorded on the error log and the error report.

An error must be evaluated to determine what category it falls into. Any error that requires a change to a controlled product or document must be reviewed by the review process (refer to QPC 006—Software Review). If the product or document is uncontrolled then the error report is sufficient to initiate a change.

5 CORRECTIVE ACTIONS

5.1 INITIAL PROCEDURE

If any error requires some level of corrective action to a document or to a system then a system amendment request must be completed (see Appendix 3); this is traceable to the error report by the error report number. The system amendment request requires details of customer name and reference, system name and reference, the proposed amendment to correct the error, and the estimated effort necessary and the resources required.

A log of all system amendment requests is maintained centrally and all have unique numbers allocated.

5.2 REVIEW

All system amendment requests to controlled documents or systems are reviewed prior to approval for the necessary work to be carried out. The review follows the procedures laid down in QPC 006—Software Review.

The review procedures cannot be circumvented for any reason and must always be applied.

Any uncontrolled documents or systems are amended via the error report which gives a sufficient level of control.

5.3 CONTROL

The control mechanism detailed in QPC 002—Configuration and Change Control are used to control the actions of any system amendment requests that are approved. Whilst a change is being processed the error report will remain open and is referenced to the system amendment request and vice versa, and also the review number.

The error report is referred to by the system amendment request form and the system amendment request refers to the error report form. The review form only refers to the system amendment request as a review cannot occur without such a form. Refer to QPC 006—Software Review for further details.

APPENDIX 1: ERROR REPORT

Error Report			
Error Number:	Date: / / Time:		Solved
Client: Reported by: Tel.: Logged by:	Client Ref.: Position: Extn.:		

Hardware: CPU model
          Device with error:
	Software: Operating system:
	          Package:		

Error type	Hardware	Operating system	Package software	User training
	User manual	Other:		

Error description:

(Description of error situation and or problem)

Problem occurred on	Terminal	Printer	CPU	Other:
	On-line	Batch	Other:	
Check	Power to CPU	Power to device	System available	

Priority of error/problem: (Urgent = 1, Important = 2, Non-urgent = 3)

Error passed to: Date: / /   Time:

First call back by: Date: / /   Time:

Solution:

(Description of solution applied or reference to system amendment request)

Referred to: Date: / /   Time:

Answer received from: Date: / /   Time:

Resolved by: Date: / /   Time:

APPENDIX 2: ERROR CONTROL LIST

Error Control List							
Error no.	*Reported*		*Client*	*Type*	*Solved*		
	Date	*Time*			*Date*	*Time*	

APPENDIX 3: SYSTEM AMENDMENT REQUEST

System Amendment Request			
System amendment request no.:		Assigned: / /	
Client name: System name: Function name: Prepared by:		Client ref.: System ref.: Function ref.: Date: / /	
Answer (Y/N)	Functional spec.	System desc.	Operational text
Amendment effects:			
Amendment details: *(Description of amendment required)*			
Review number:		Date reviewed: / /	

Cost details:			Accept/Reject* (* Delete as necessary)
Resource......................	Estimated		
	Days	Cost (£)	Authorisation .... Signature: Printed name: Position:
			Requirement date: / /
			Functional spec. altered: / /
			System description altered: / /
Total cost(s):			Operational text altered: / /

428

COMPANY C
QUALITY PROCEDURE QPC 018
QUALITY RECORDS

Compiled by...

Checked by...

Approved by...

Date...

CHANGE HISTORY PAGE

Document status	Date issued	Number of pages	Changed pages	Change/ defect no.
XX	XX	XX	XX	XX

CONTENTS LIST

1 INTRODUCTION

1.1 PURPOSE OF DOCUMENT

To ensure that the records produced by the quality system are kept and maintained, especially where the records have a contractual aspect. The period of retention is also specified.

1.2 SCOPE

All records produced by the quality system.

1.3 TERMS AND ABBREVIATIONS

None.

2 APPLICABLE DOCUMENTS AND REFERENCES

QMC 001—Quality Manual
All QPC/QPS procedures

3 QUALITY RECORDS

The most important quality records include the following.

3.1 IDENTIFICATION REPORT

The identification report will be retained in the customer file for 3 years, after which it will be archived for a further 2 years.

3.2 FEASIBILITY STUDY

The feasibility documents will be retained in the customer file for 3 years, after which they will be archived for a further 2 years.

3.3 PROPOSAL AND CONTRACT

The proposal will be retained in the customer file for 3 years, after which it will be archived for a further 2 years.

The contract itself may need to be retained for a further period and the contract conditions should be consulted in order to determine this.

3.4 SYSTEM DESCRIPTION

This will be retained in the customer file for 3 years, after which it will be archived for a further 2 years.

3.5 TEST PLAN

This includes test plans, test results, test programs and error reports. They will be kept in the project file for 1 year after completion of the project and then archived for a further 2 years.

3.6 REVIEW FORMS

All review forms will be retained in a file held by the technical director. They will be kept in the project file for 1 year after completion of the project and then archived for a further 2 years.

3.7 AUDIT REPORTS

All internal audit reports will be kept in an audit file held by the technical director. They will be kept in the project file for 1 year after completion of the project and then archived for a further 2 years.

3.8 ERROR REPORTS

Error reports, from all sources, and error control lists will be kept in an error report file by the technical director. They will be kept in the project file for 1 year after completion of the project and then archived for a further 2 years.

3.9 SYSTEM AMENDMENT REQUESTS

These will be kept in the project file and a copy retained by the technical director. The requests will be kept in the project file for 1 year after completion of the project and then archived for a further 2 years.

3.10 PURCHASING AND SUPPLIER DOCUMENTS

Assessment records, goods received notes, questionnaires and supplier quality assurance rating forms shall be retained for as long as Company C is dealing with that supplier. Thereafter documents will be retained in a supplier file for 3 years, after which they will be archived for a further 2 years.

3.11 OTHER RECORDS

These will be retained at the discretion of the technical director.

COMPANY C
QUALITY PROCEDURE QPC 019
INTERNAL QUALITY AUDIT

Compiled by..

Checked by..

Approved by ....

Date..

CHANGE HISTORY PAGE

Document status	*Date issued*	*Number of pages*	*Changed pages*	*Change/ defect no.*
XX	XX	XX	XX	XX

CONTENTS LIST

1 INTRODUCTION

1.1 PURPOSE OF DOCUMENT

The purpose of this procedure is to provide guidance on carrying out internal audits at Company C. It is the purpose of the audit to ensure that standards and procedures are adhered to.

1.2 SCOPE

The scope is internal to Company C. If the need to audit suppliers arises this document will be expanded accordingly.

1.3 TERMS AND ABBREVIATIONS

None.

2 APPLICABLE DOCUMENTS AND REFERENCES

BS 5750 (Part 1) 1987
QPC 008—Documentation Format and Standard
QPC 004—Software Development
QPC 006—Software Review
QPC 005—Software Design Standards
QPC 002—Configuration and Change Control
QPS 007—Subcontract Equipment and Software
QPC 015—Test
QPC 017—Non-conformance, Corrective Action and Records

3 GENERAL AUDIT REQUIREMENTS

Audits will be carried out at the instigation of the technical director in consultation with the project manager responsible for the project or product.

3.1 OBJECTIVES OF THE AUDIT

The objective of the audit is to ensure that procedures are being adhered to and that controls exist so that quality is maintained. Audits shall be carried out at regular and predefined intervals and shall cover the following:

—Establish that there is adequate control over design for a product which is under development (QPC 004).
—Establish that procedures are being used for the documentation and production of a product (QPC 008).
—Ensure that procedures are being periodically reviewed (QPC 006).
—Ensure that software design standards are being followed (QPC 005).
—Establish that release acceptance procedures are being followed and that strict update configuration control is being applied (QPC 002).
—Establish that error handling procedures are being followed (QPC 017).
—Ensure that test procedures are being followed (QPC 015).
—Establish that any products which are externally controlled comply with the requirements for subcontracted software (QPS 007).

3.2 IMPLEMENTATION OF THE AUDIT

The technical director will maintain an internal audit file, which will contain:

—planned future audits at defined intervals (e.g. monthly)
—audit records
—outcome of remedial action resulting from previous audits

The audits will be planned ahead by the technical director, who will maintain a file of forward plans which address the implementation of the quality manual and will describe the features of the system that are to be audited.

The audit must seek objective evidence that the specific procedures are being implemented. Formal records will be kept and will include:

—the audit report
—reports of deficiencies
—remedial actions with target dates

3.3 THE INTERNAL AUDIT REPORT

At the end of the audit a report will be prepared by the technical director detailing:

—those involved and their roles;

437

—an overview of the recommendations (one page maximum);
—the checklists used in the audit;
—details of any concessions;
—corrective actions and target dates.

3.4 FREQUENCY OF AUDITS

The audit should be a regular feature of all work undertaken by Company C. Typically there should be one audit per month on some aspect of the quality manual and each project should receive at least one audit.

COMPANY C
QUALITY PROCEDURE QPC 021
SERVICING

Compiled by..

Checked by..

Approved by..

Date..

CHANGE HISTORY PAGE

Document status	*Date issued*	*Number of pages*	*Changed pages*	*Change/ defect no.*
XX	XX	XX	XX	XX

CONTENTS LIST

1 INTRODUCTION

1.1 PURPOSE OF DOCUMENT

This procedure is concerned with providing the necessary detail for successful post-delivery service of a project. It describes the general responsibilities and actions involved.

1.2 SCOPE

All products and services supplied by Company C.

1.3 TERMS AND ABBREVIATIONS

None.

2 APPLICABLE DOCUMENTS AND REFERENCES

QPC 001—Contract Review
QPC 002—Configuration and Change Control
QPC 015—Test
QPC 017—Non-conformance, Corrective Action and Records
QPC 022—Statistical Techniques

3 ERROR LOGGING

Errors will be logged as described in QPC 017.

3.1 ERROR REPORTS

The error details logged are described in QPC 017 along with the format. In addition, the time taken to diagnose and correct the problem will be recorded, as will the effects of the defect on the system in terms of symptoms and down-time.

3.2 ERROR REPORT SUMMARIES

Periodic error report summaries will be prepared and trend analysis will be carried out by the technical director. These analyses will form the basis of management decisions relating to remedial action arising from defects. QPC 022 describes some methods of trend analysis.

4 CONFIGURATION MANAGEMENT

All error reports will be subject to configuration control as described in QPC 002. Corrective actions will only be authorised after a system amendment request has been raised and authorised. An impact analysis should be undertaken to evaluate any special processes or requirements needed to implement the amendment.

4.1 EVALUATION

The error reports will be subject to detailed evaluation as described in QPC 017.

4.2 INSTALLATION

If, as a result of the evaluation, revised installation procedures are indicated for the applications software or for the operating system then changes will be put in hand.

4.3 CONFIGURATION AUDITS

Regular audits will be carried out on service and warranty records. These will be supervised by the technical director. Any discrepancies will be subject to corrective action and a planned timetable will be laid down.

5 CONTRACT REVIEW

All service and warranty contracts will be reviewed annually to ensure that the conditions continue to be applicable.

6 TRAINED PERSONNEL

Suitable trained personnel will be made available for investigating and evaluating error reports, and a design authority will be identified in each instance to manage the evaluation.

Resource levels will be maintained such that support can be provided.

COMPANY C

QUALITY PROCEDURE QPC 022

STATISTICAL TECHNIQUES IN
SOFTWARE DEVELOPMENT

Compiled by...

Checked by...

Approved by...

Date...

CHANGE HISTORY PAGE

Document status	Date issued	Number of pages	Changed pages	Change/ defect no.
XX	XX	XX	XX	XX

CONTENTS LIST

1 INTRODUCTION

1.1 PURPOSE OF DOCUMENT

The purpose of this procedure is to provide an integrated approach to software data collection, analysis and interpretation using Pareto-type statistical techniques which are relevant to software development.

1.2 SCOPE

All software and firmware development can be categorised after appropriate inspection or review. Error types can be derived and trends predicted.

1.3 TERMS AND ABBREVIATIONS

None.

2 APPLICABLE DOCUMENTS AND REFERENCES

QPC 002—Configuration and Change Control
QPC 006—Software Review
QPC 017—Non-conformance, Corrective Action and Records

3 STATISTICAL TECHNIQUES FOR SOFTWARE

Statistical methods are widely used in areas such as market analysis, product design, engineering, manufacturing processes and sampling inspection. Data is collected for well-established classifications. 'Metrics' (e.g. failure rate, availability, life-time, dimension) are derived. These measures, and their distributions, bear upon such areas as:

—design of experiments
—risk analysis
—regression analysis
—statistical process control

These methods and their applications are usually hardware related where conformance can be demonstrated by well-established measurements and methods.

For software- and firmware-based products, however, many of these statistical methods are not satisfactory. Sampling is inappropriate because software lacks visibility and it is not clear what should be measured or by what means. Testing alone, on the other hand, is never complete because in practice it only demonstrates a proportion of the possible execution paths within the code.

Applying the Pareto principle to software development can provide a sound basis for collecting software statistical data and for analysing it. Error types and frequencies can be derived and appropriate action taken.

The Pareto technique can be applied to:

—defect identification
—code inspections
—error types
—error frequencies

4 THE PARETO PRINCIPLE

The basic principle is to concentrate on the vital few rather than the trivial many. It is frequently the case that a small proportion of failures often constitute the major contribution to the overall effect. This phenomenon is known as the Pareto principle (after the Italian economist).

It is finding increasing favour as a means of reviewing software failures.

5 DEFECT IDENTIFICATION

5.1 FAILURE CAUSES

The types of defects need to be established by category. The hypothetical example shown in Table 1, based on work by Thayer, on real time systems using Fortran and assembler languages shows the percentage breakdown of defects into identifiable types.

Logic and data handling errors rank first and second, and were thus chosen for further breakdown and analysis into type, as shown in Table 2.

The examples in Tables 1 and 2 show that the most common logic error was missing logic, and that the most frequent data handling problem related to incorrect initialisation. The principle of analysing failure causes is important and data should be collected in this manner.

446

TABLE 1

Error category	Project 1 (%)	Project 2 (%)	Applications software (%)	Simulator software (%)	Operating system (%)
Computational	10	0	15	20	5
Logic	20	35	20	20	35
Data input	15	10	5	15	10
Data handling	20	25	10	15	20
Interface	20	25	10	5	5
Data definition	0	3	5	10	5
Database	5	2	25	15	5
Other	10	0	10	5	15

TABLE 2

	Applications software (%)	Simulator software (%)	Operating system (%)
Logic errors			
Logic	5	10	5
Incorrect operand in logical expression	20	5	5
Logic activity out of sequence	25	30	10
Wrong variable being checked	5	5	15
Missing logic or condition test	45	50	65
Data handling			
Data handling error	10	25	10
Data initialisation not done	5	10	15
Data initialisation done improperly	20	10	45
Variable used as a flag or index not set properly	20	5	25
Variable referred to by wrong name	5	20	0
Bit manipulation done incorrectly	10	0	0
Incorrect variable type	5	10	0
Data packing/unpacking error	10	5	0
Subscripting error	15	15	5

TABLE 3

Error categories	% of total code change errors	Probable sources	
		% Design	% Code
Computation	15	90	10
Logic	25	80	20
I/O	15	25	75
Data handling	15	25	75
System support S/W	0		
Configuration	5	25	75
Routine interfaces	15	55	45
Software interface	2	70	30
Tape interface	0	90	10
Globals	5	80	20
Compool definition	0	65	35
Recurrent	1		
Documentation	0		
Requirements compliance	0	90	10
Unidentified	1		
Operator	0		
Questions	1		
TOTAL		62	38

5.2 BASIC SOURCES OF DEFECTS

Defects may also be categorised by source and a further example is given in Table 3.

Table 3 shows that twice as many errors are caused during design as arise from coding. The left-hand column also suggests input/output, logic and data handling as the main sources of difficulty.

5.3 MODULE DEFECTS

A further area for investigation is software modules. A method is needed to identify modules whose defects will have an impact on the rest of the system. A module can be highly 'coupled' (that is to say, has many interfaces with other modules) with the rest of the system as a result of the parameters it passes on or the global data which it accesses. In these cases one should seek to reduce these interfaces since modifications to such modules are more dangerous than to others. The defect analysis should seek to identify these problems.

6 INSPECTIONS

This Pareto approach had long been applied to assembly line production and can equally be applied to code inspections. Participants should be familiar with the ranked distributions of failures from previous inspections. They may well be in the format shown in the previous section. This concentrates the attention on the vital few categories which are likely to cover the majority of defects.

7 STATISTICAL TECHNIQUES

The use of histograms is useful in presenting a graphical view of the distribution of errors. The following is an example of a histogram.

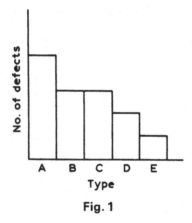

Fig. 1

8 SUMMARY

- The Pareto approach should be adopted in the recording and analysis of software errors.
- Firstly, arrange the items in order of importance.
- Secondly, identify the vital few categories which contain the majority of defects.
- Use graphical (histogram) methods of presentation in reports.

Sources of Information

QUALITY, STANDARDS AND FINANCE

For up-to-date information on publications, the following (UK) sources are recommended. Financial support is, in some cases, available through the DTI and PERA (see below) for small companies wishing to introduce quality management systems into their company.

- The British Computer Society (BCS)
  13 Mansfield Street
  London W1E 2YZ

- The British Standards Institution (BSI)
  PO Box 375
  Milton Keynes MK14 6LL

- Computer Services Association (CSA)
  Hanover House
  73/74 High Holborn
  London WC1V 6LE

- Electronic Engineering Association (EEA)
  Leicester House
  8 Leicester Street
  London WC2H 7BN

- Confederation of British Industry (CBI)
  Centre Point
  103 New Oxford Street
  London WC1A 1DU

● The Institution of Electrical Engineers (IEE)
Savoy Place
London WC2R 0BL

● The Institution of Electrical and Electronic Engineers' Computer Society
(IEEE)
All publications are available at the IEE Library, Savoy Place, London
WC2R 0BL

● Health and Safety Executive (HSE)
HMSO Bookshops

● Lloyd's Register
Norfolk House
Wellesley Road
Croydon CR9 2DT

● Ministry of Defence (MOD)
Directorate of Standardisation
First Avenue House
High Holborn
London WC1V 6LL

● Director General Defence Quality Assurance
Woolwich Arsenal
London SE18 6TD

● North Atlantic Treaty Organisation (NATO)
Publications are available from MOD as above

● The National Computing Centre (NCC)
Oxford Road
Manchester M1 7ED

● National Physical Laboratory (NPL)
Queen's Road
Teddington
Middlesex TW11 0LW

• The Institute of Quality Assurance
  10 Grosvenor Gardens
  London SW1W 0DQ

• The Safety and Reliability Society
  Clayton House
  59 Piccadilly
  Manchester M1 2AQ

• Systems Reliability Service (SRS)
  United Kingdom Atomic Energy Authority
  Culcheth
  Warrington WA3 4NE

• Production Engineering Research Association (PERA)
  Melton Mowbray
  Leicestershire LE13 0PB

• Department of Trade and Industry (DTI)
  For details on the DTI's Quality and Enterprise Initiatives and 1992 Single European Market contact a local DTI office:

  —DTI East (Cambridge)
   Building A
   Westbrook Research Centre
   Milton Road
   Cambridge CB4 1YG
   Tel: (0223) 461939
   Telex: 81582

  —DTI East Midlands
   Severns House
   20 Middle Pavement
   Nottingham NG1 7DW
   Tel: (0602) 506181
   Telex: 37143 DTINOTG

  —DTI North-East
   Stanegate House
   2 Groat Market
   Newcastle-upon-Tyne NE1 1YN
   Tel: 091-235 7270
   Telex: 53178 DTITYNG

—DTI North-West (Liverpool)
Graeme House
Derby Square
Liverpool L2 7UP
Tel: 051-227 4111
Telex: 627647 DTIPLG

—DTI North-West (Manchester)
Sunley Tower
Piccadilly Plaza
Manchester M1 4BA
Tel: 061-838 5227
Telex: 667104 DTIMCHRG

—DTI South-East (London)
Bridge Place
88–89 Eccleston Square
London SW1V 1PT
Tel: 071-215 0576
Telex: 297124 SEREXG

—DTI South-East (Reading)
40 Caversham Road
Reading
Berkshire RG1 7EB
Tel: (0734) 395600
Telex: 847799

—DTI South-East (Reigate)
Douglas House
London Road
Reigate RH2 9QP
Tel: (0737) 226900
Telex: 918364

—DTI South-West
The Pithay
Bristol BS1 2PB
Tel: (0272) 272666
Telex: 44214 DTIBTLG

—DTI West Midlands
Ladywood House
Stephenson Street
Birmingham B2 4DT
Tel: 021-631 6181
Telex: 337919 DTIBHAMG

—DTI Yorkshire and Humberside
Priestley House
3–5 Park Row
Leeds LS1 5LF
Tel: (0532) 443171
Telex: 557925 DTILDSG

In Scotland, Wales and Northern Ireland you should contact:

—Scottish Office, Industry Department for Scotland
Alhambra House
45 Waterloo Street
Glasgow G2 6AT
Tel: 041-248 2855
Telex: 777883 IDS GWG

—Industrial Development Board
IDB House
64 Chichester Street
Belfast BT1 4JX
Tel: (0232) 233233
Telex: 747152 DEC DEV G

—Welsh Office Industry Department
Cathays Park
Cardiff CF1 3NQ
Tel: (0222) 825111
Telex: 498228 WOCARD G

Bibliography

Department of Trade and Industry (DTI) now publish many booklets on quality assurance, control and management. These are an excellent introduction to the subject. Details are available from:

IBIS Information Services Limited, Waterside, Lowbell Lane, London Colney, St Albans AL2 1DX (Tel: 0727 24777)

Services Ltd, Quality and Reliability House, 82 Trent Boulevard, West Bridgford, Nottingham NG2 5BL (Tel: 0602 455285)

Babich, W. A. (1986). *Software Configuration Management.* Addison-Wesley.
Beizer, B. (1983). *Software Testing Techniques.* Van Nostrand Reinhold.
Beizer, B. (1984). *System Testing and Quality Assurance.* Van Nostrand Reinhold.
Bernstein, D. (1990). *Working for Customers.* CBI, London.
Buckle, J. K. (1982). *Software Configuration Management.* Macmillan, London.
Collard, R. (1989). *Total Quality.* Institute of Personnel Management, London.
Crosby, P. B. (1979). *Quality is Free.* McGraw-Hill, New York.
Crosby, P. B. (1984). *Quality Without Tears.* McGraw-Hill, New York.
Deming, W. E. (1982). *Quality, Productivity and Competitive Position.* Center for Advanced Engineering Study, MIT Press, Massachusetts, USA.
Deming, W. E. (1986). *Out of the Crisis.* Center for Advanced Engineering Study, MIT Press, Massachusetts, USA.
Deutch, M. S. (1982). *Software Verification and Validation.* Prentice-Hall.
Evans, M. & Marcinack, J. (1987). *Software Quality Assurance and Management.* John Wiley & Son, London.
Fiegenbaum, A. V. (1983). *Total Quality Control.* McGraw-Hill, New York.
Groocock, J. (1986). *The Chain of Quality.* John Wiley & Sons, New York.
Juran, J. M. (1964). *Managerial Breakthrough.* McGraw-Hill, New York.
Juran, J. M. (1988). *Juran on Planning for Quality.* The Free Press (Macmillan), New York.
Juran, J. M. & Gryna, F. M. (1980). *Quality Planning and Analysis.* McGraw-Hill, New York.

Juran, J. M., Gryna, F. M. & Bingham, R. S. (1987). *Quality Control Handbook*. McGraw-Hill, New York.

Manns, T. & Coleman, M. (1988). *Software Quality Assurance*. Macmillan Educational, London.

Musa, J. (1987). *Software Reliability, Measurement, Prediction and Application*. McGraw-Hill, New York.

Oakland, J. S. (1989). *Total Quality Management*. Heinemann, Oxford.

Quirk, W. J. (ed.) (1985). *Verification and Validation of Real-time Software*. Springer-Verlag.

Schulmeyer, G. G. & McManus, J. I. (eds) (1987). *Handbook of Software Quality Assurance*. Van Nostrand Reinhold, Wokingham, Berkshire, UK.

Smith, D. J. (1988). Statistics Workshop (an introduction to statistical sampling and inference by means of practical simulations). Technis. *Available from IQA, 10 Grosvenor Gardens, London SW1W 0DQ*.

Smith, D. J. & Locke, D. (eds) (1990). *The Handbook of Quality Management*. Gower Press, Aldershot.

Smith, D. J. & Wood, K. B. (1987). *Engineering Quality Software*. Elsevier, London.

Index

Printed in the United States
By Bookmasters